How We Do It

How We Do It

The Evolution and Future of Human Reproduction

ROBERT MARTIN

BASIC BOOKS
A Member of the Perseus Books Group
New York

Copyright © 2013 by Robert Martin

Published by Basic Books,
A Member of the Perseus Books Group

All rights reserved. Printed in the United States of America.
No part of this book may be reproduced in any manner whatsoever without written permission except in the case of brief quotations embodied in critical articles and reviews. For information, address Basic Books, 250 West 57th Street, 15th Floor, New York, NY 10107-1307.

Books published by Basic Books are available at special discounts for bulk purchases in the United States by corporations, institutions, and other organizations. For more information, please contact the Special Markets Department at the Perseus Books Group, 2300 Chestnut Street, Suite 200, Philadelphia, PA 19103, or call (800) 810-4145, ext. 5000, or e-mail special.markets@perseusbooks.com.

A CIP catalog record for this book is available from the Library of Congress.

ISBN: 978-0-465-03015-6 (hardcover)
ISBN: 978-0-465-03784-1 (e-book)

10 9 8 7 6 5 4 3 2 1

I dedicate this book to Anne Elise—
natural mother par excellence and my best friend

CONTENTS

INTRODUCTION

Alone with their baby for the first time, a new mother and father are flooded with novel emotions and a whole new kind of love. Although they cannot know what they will experience over the weeks and months ahead, they have now experienced firsthand the miracle of life. For them, this moment is one of a kind, a unique event in the history of humankind. Yet to reach this point took not nine months but millions of prior generations. What the mother has undergone and how the pregnancy and mothering will change her, what the father has contributed, and how this tiny, helpless infant will gradually develop into a full-grown adult belong to a story much bigger than theirs. It is the natural history of how human beings reproduce.

Human reproduction has an extensive natural history, and biological adaptations with that kind of pedigree demand our attention. However, the evolutionary background to human reproduction rarely has been examined in depth. This is odd, as successful breeding is the key to evolution.

Furthermore, despite the common assumption that natural is better when it comes to having and raising children, few have thought to ask what is really natural—how did we evolve to reproduce and raise children? In order to answer this question, we must peer millions of years into the past. The features of our reproductive biology evolved in our vertebrate and primate forebears, including the basic anatomy of our reproductive organs, internal fertilization, breast-feeding, and the practice of toting babies around. By understanding how—and in what environment—these features evolved, we can make reproductive decisions that ensure our health and well-being and that of our offspring.

But this book is not about returning to a primitive lifestyle. In today's world it would be absurd to try to return to the kind of existence led by our primate forerunners or even our gatherer-hunter ancestors. Rather, it will provide much-needed context for our current reproductive behaviors in order to dispel harmful notions, promote practices and technologies that reflect a deep understanding of natural human biology (e.g., the birth control pill), and, most of all, put minds at ease about such fraught topics as breast-feeding duration, different forms of birth control, and attachment parenting. The scientific mysteries explored will, I hope, lead readers to better decisions on their reproductive journeys. Ultimately, my goal is to enhance the richness and naturalness of the reproductive experience by connecting it with the entire history of *Homo sapiens*, that remarkably equipped primate.

EVERYTHING BEGAN WITH SEX, so that is where the story begins. In Chapter 1 I discuss the evolution of human sex cells and explain why sperms and eggs are so different in size. Why does it take a quarter of a billion sperms to fertilize one human egg? We really need an answer to this question, as there is mounting evidence for a worldwide decline in sperm counts, with rising levels of bisphenol A as a prime suspect for its cause. Chapter 2 is about cycles and seasonality. Why do women, apes, and monkeys menstruate, while most other mammals do not? Are women fertile only during a few days in each menstrual cycle? I explore the implications of seasonal patterns in conception and birth, especially regarding the conflict between our biological clocks and the modern age of electric light.

Chapters 3 and 4 take us from mating to conception. What is natural in human mating—monogamy, harems, or promiscuity? Are our reproductive systems adapted for competition between sperms from different men? Turning to pregnancy and birth, I look at possible evolutionary reasons for the dreaded morning sickness. Did nausea and vomiting evolve to protect the fetus against perils of the mother's diet? I also explore the mystery of why human birth is uniquely challenging, making help from midwives almost indispensable.

We move on to development in Chapter 5, which focuses on the vital topic of the human brain and how its uniquely large size has been a driver for many adaptations in the way we reproduce and care for our young. Why are human babies born in a comparatively immature state, requiring intensive caregiving? Chapter 6 is a natural history of suckling, in which I suggest that breast-feeding—while certainly not an option in all households today—provides benefits for both baby and mother that proponents of bottle-feeding have yet to match successfully. Chapter 7 opens up to a broader discussion of child care, including a discussion of infant carriage, an 80-million-year-old primate adaptation that laid the foundation for suckling on demand and mother-infant bonding. In Chapter 8, I close with a discussion of how we apply scientific understanding of our reproduction to constrain it through contraception or promote it through assisted reproduction.

WHENEVER I GIVE a public lecture on human evolution, during question time somebody in the audience is likely to ask me what I can say about the future. Evolutionary biology is a largely historical science, and we cannot reliably predict how evolution will proceed. But many medical interventions have reduced or even arrested the action of natural selection. Genetic predispositions that would have been purged by negative selection in the past now slip through the net. For instance, we deliberately bypass natural selection when we combat infertility with assisted reproduction. Individuals with a genetic predisposition for infertility may pass this on to offspring that would not have been born naturally. Fertilizing an egg with a single sperm in the laboratory, for instance, circumvents natural filtering mechanisms. Indeed, reproductive technology has developed to the point where most complications can be overcome, whether it's infertility, premature birth, or otherwise compromised deliveries. And while we seemingly have found ways to solve many reproductive challenges—to the relief and joy of parents everywhere—delight has overshadowed basic questions about long-term consequences.

This book has roots descending deep into the past and took more than fifty years to germinate from its initial seed. But the core subject of evolution has far, far deeper roots, stretching back more than 3 billion years to the beginnings of life on Earth. So perhaps I have some excuse for taking so long. At any rate, I present here what I hope is a comprehensive examination of the subject, from sex cells to weaning. It is a full reconstruction of our evolutionary past, aimed at providing a better basis for understanding our present and future.

My goal is not to convince you that we have firm answers to any of the questions I have raised, but rather to make you aware of how far we have come in understanding the natural background to human reproduction in the 150 years since the science of evolution—and the exploration of human origins—was launched. One of the key lessons that I have learned from studying the natural world is enormous respect for complex systems that have evolved over millions or even billions of years. With this book, I would like to share that powerful feeling with fellow parents everywhere, past, present, and future.

CHAPTER 1

Sperms and Eggs

"Where do babies come from?" Many parents sidestep a truthful answer to this innocent childhood question, instead mumbling about storks or gooseberry bushes. But until only three hundred years ago *any* answer would have been a fairy tale. In antiquity, pregnancy was thought to result from mixing semen with menstrual blood. At first contact a few modern populations, such as the Arunta tribe of Australia and the Trobriand Islanders of Oceania, reportedly still had not connected copulation to pregnancy, although it later emerged that Trobrianders also believed that mixing semen with menstrual blood led to pregnancy. We don't know how and when people first recognized the link between sex and pregnancy, but it was probably not far back in prehistory. The domestication of mammals, starting around 10,000 years ago, must have yielded telling clues, with a basic grasp of reproductive biology being one early by-product of animal husbandry. Procedures such as castration, widely used to reduce aggression in male mammals, would have led to additional insights. Yet conception and birth are nine months apart in humans, so a major breakthrough was needed to recognize the link. Even then, fanciful notions regarding reproductive mechanisms persisted long afterward.

The discovery of sex cells—the essential elements of conception—did not occur until the seventeenth century, when Dutch tradesman-scientist Anton van Leeuwenhoek first observed sperms with a microscope in 1667.

At first he mistook them, thinking they were tiny parasites that had contaminated the seminal fluid. Mammal sperms are invisible to the naked eye, so until van Leeuwenhoek and other pioneers developed microscopes strong enough to see them, it would have been impossible to understand their part in conception. Although mammal eggs are thirty times wider than sperms and just visible to the naked eye (about the size of the period at the end of this sentence), they were discovered much later. German biologist Karl Ernst von Baer, the founder of embryology, published the first report on eggs from humans and sundry other mammals in 1827. In fact, it was von Baer who, that same year, coined the term "spermatozoon" from the Greek *sperma*, "seed," and *zōon*, "living thing."

Furthermore, it took some time before anyone recognized the purpose of the sex cell. For centuries, people believed that organisms emerged directly from lifeless matter through spontaneous generation—for example, maggots arising from rotting flesh. It was not until the eighteenth century that, through a series of eccentric experiments, Italian priest and natural scientist Lazzaro Spallanzani provided the first clear evidence against spontaneous generation and proved that sperms were needed to fertilize eggs. In the 1760s, he fitted male frogs with tight-fitting taffeta pants to show that eggs would not develop into tadpoles unless sperms were shed into the surrounding water. This, it seems, was the first experimental demonstration of barrier contraception. Fertilization occurred only when Spallanzani retrieved semen that had been discharged into the pants and painted it onto the eggs, demonstrating the principle of artificial insemination as well.

The discovery of the mechanisms of sexual reproduction thus occurred quite recently in human history. Yet while we know a lot about making babies, we don't know very much about why we make them this way. At all levels of complexity questions arise regarding human sex and sex cells. To start with, it is not obvious why sexual reproduction occurs at all. Surely self-replication would be simpler, more reliable, and less messy. Furthermore, why do humans produce distinctive male and female sex cells?

For clues to these questions, we need to broaden our view and look at how and why sexual reproduction evolved in our ancestors. We are used to comparing ourselves with other primates, but some of our more basic reproductive features can be traced to previous stages of animal evolution that stretch far back into the geological past. In this chapter I examine some very early stages in the evolution of life in our quest to reach a full understanding of human sex cells.

A BASIC QUESTION THAT HAS long puzzled evolutionary biologists is why sex exists at all. Sexual reproduction is widespread, especially among organisms most familiar to us, yet it appears at first to present an evolutionary enigma. If an organism replicates itself without sex—for example, by dividing or budding—all offspring will have the same genes. By contrast, when an organism breeds sexually, each offspring has a mixture of genetic material, half from one parent and half from the other. So unless a major benefit offsets the cost of producing offspring with 50 percent of their genes from another individual, sexual reproduction should be at a severe disadvantage under natural selection. But because sexual reproduction is so common among animal species, you might expect that it must have some strong and systematic advantage. And you would be right. The most likely explanation for the prevalence of sexual reproduction is that the fusion of two sex cells from different individuals generates variability by mixing genes. Variety is not just the spice of life but its very essence. Natural selection simply cannot work without it; without variation, evolutionary change is impossible. Breeding without sex, also known as cloning, has a significant drawback: Without fertilization to ensure a chance reshuffling of genes, variation can arise only as mutations accumulate—slowly. The question is whether the advantage of variability supplied by sexual reproduction is enough to offset the disadvantage of producing offspring with half of their genes from a sexual partner. Various experiments with simple organisms have indicated that sexual reproduction does, indeed, have an edge. It permits a more rapid response to new selection pressures under changed environmental conditions. In other words, it allows species to adapt more readily to external changes—and humans are nothing if not adaptable.

This reshuffling of genes is important for an individual's ability to cope with threats from other organisms. For instance, if natural selection favors predators that can run faster to catch their prey, it will also favor the evolution of faster-running prey to improve their chances of escape. Evolutionary biologist Leigh Van Valen famously recognized this ongoing arms race between interacting species in what is now widely known as the Red Queen principle. This name comes from Lewis Carroll's iconic children's book *Through the Looking-Glass*, in which the Red Queen tells Alice: "It takes all the running you can do, to keep in the same place." In other words, a species often must evolve rapidly simply in order to maintain its niche in the natural environment, and sexual reproduction allows it to do so. Simple enough, but Van Valen—an iconic figure at the University of Chicago who was adored by students—encountered resistance from established journals when he tried to publish his paper announcing the Red Queen principle. In response he launched his own journal, *Evolutionary Theory*, and published his seminal idea in a paper entitled "A New Evolutionary Law" in the first issue in 1973. That once "unpublishable" notion, rebuffed in the lottery of peer review, is now one of the established principles of evolutionary biology.

However it evolved, sexual reproduction is now dominant in living organisms and is virtually universal among animals with backbones: fish, amphibians, reptiles, birds, mammals. It is therefore safe to conclude that it was already established in a common ancestor around 500 million years ago. Yet with the evolution of land-living reptiles, birds, and mammals arose an important innovation: internal fertilization. Fish and amphibians can simply release eggs and sperms into the water around them, and fertilization takes place externally. By contrast, the common ancestor of land-living vertebrates developed internal fertilization as a crucial adaptation to life away from water. Special adaptations to protect developing offspring also accompanied the shift to dry land. Most reptiles, birds, and monotremes (the platypus and echidnas) lay eggs with shells that resist drying. Marsupials and placentals went one step further, developing internal development leading to live birth (vivipary).

To UNDERSTAND THE EVOLUTION of sex cells, though, we must reach much farther back in time. A fundamental early transition in evolution, almost 1.5 billion years ago, occurred with the rise of single-cell organisms with a proper nucleus. Inside the nucleus were chromosomes that carried most of an organism's genetic material, in the form of deoxyribonucleic acid, or DNA. In an unusual development accompanying the evolution of cells with a nucleus, bacteria that were once free-living became permanent residents in the cell substance surrounding the nucleus—with far-reaching implications for the evolution of all animals and plants.

Over time, the resident bacteria transformed into mitochondria, which are often described as cell powerhouses because they play a direct part in energy turnover. Intriguingly, DNA comparisons have revealed that mitochondria are quite closely related to the *Rickettsia* bacterium that causes epidemic typhus in humans. We do not know whether mitochondria originally had some mutually profitable arrangement with the host cell or whether they were simply captured and enslaved. What we do know is that the host cell provided them with accommodation, food, and protection in exchange for energy production round the clock. All the bacteria lost was their freedom.

So mitochondria evolved as stripped-down bacteria. As an inheritance from its free-living ancestry, each mitochondrion retains a few ring-shaped gene sets. These rings are passed on separately from genes carried by chromosomes in the host cell's nucleus. As a result, all nucleus-containing cells, including those of humans, have two separate gene sets: a nuclear genome and a mitochondrial genome. In both cases, most individual genes prescribe the composition of specific proteins. However, the gene sets of mitochondria have gradually diminished over the hundreds of millions of years since they became permanent residents of cells with a nucleus. In mammals, for example, each mitochondrion now retains only thirteen protein-coding genes from an original set of about fifty. Proteins produced by those thirteen genes are all enzymes connected with energy turnover, which is why mitochondria are known as cell powerhouses.

Another milestone for both animal and plant reproduction was the evolution of bodies containing multiple cells. This transition occurred more than 600 million years ago. The presence of many cells in an individual body soon led to division of labor: different cells specialized to perform different functions. A human body contains about two hundred different kinds of cells, including separate sex cells for reproduction. In some organisms, such as backboned animals, a separate, everlasting lineage of founder cells (the germ line) developed to produce sex cells. A distinct germ line maintains continuity from generation to generation, regardless of what happens to the rest of the body. In this way, a body can be seen as a temporary structure that enables the germ line to fulfill its fundamental task of reproduction in each generation. Evolutionary biology therefore has a convincing answer to that age-old conundrum: "Which came first, the chicken or the egg?" The egg, of course, came first. Chickens, like people, evolved so that one egg could lead to another.

In a further step connected with the origin of germ lines, most but not all multicellular organisms also developed a sexual division of labor, with females producing eggs and males producing sperms. In some species, single individuals produce both sperms and eggs and are known as hermaphrodites, with a nod to the Greek myth of Hermaphroditus. Humans, on the other hand, follow the standard model, with women producing eggs and men sperms. From this fundamental difference stem all of the physiological and anatomical variations between the sexes.

WE TAKE IT FOR GRANTED that sperms are small and churned out in large numbers, while eggs are substantially larger and frugally produced. A man produces a multitude of tiny sperms, the smallest cells in his body, while in each cycle a woman usually releases a single egg, the largest cell in hers. The average human ejaculate, amounting to just over half a teaspoon, contains about a quarter of a billion sperms. And here is one of the lasting mysteries of human reproduction: Why are hundreds of millions of sperms needed if only one will ultimately fertilize each egg? I rather like the dismissive response: "Because not one of them will ever stop to ask the way." In all seriousness, though, it is not clear why the numbers of sperms and eggs produced differ so enormously. After all, sexual reproduction presumably started with two similarly sized single-cell organisms fusing and then dividing. Why, then, do many-celled organisms typically produce a multitude of small sperms but far fewer, larger eggs? This problem, which has long puzzled evolutionary biologists, still awaits satisfactory resolution. Natural selection may favor the production of hordes of sperms by males because it permits a genetic lottery at fertilization. But production of many small sex cells by both sexes may make it difficult for the cells to meet up for fertilization. It could be simply that the optimal solution to this biological "problem" may be a large egg serving as a sitting target for a multitude of tiny sperms. Moreover, the number of offspring produced in each batch—for example, the litter size in mammals—is also under selection. Thus the production of small numbers of eggs may provide an effective means of regulating breeding output, which will be subject to natural selection exerted by available resources.

However it evolved, successful human fertilization definitely needs large numbers of sperms. Research by infertility clinics in the 1950s revealed that men are commonly infertile if total sperm counts in an ejaculate remain persistently below 70 million. But above a certain level, sperm counts and fertility are not tightly linked. A 1953 study by medical endocrinologist Edward Tyler showed that the rate of conception improved progressively as sperm counts increased from 70 million up to 200 million per ejaculate, but the conception rate then leveled off. Recent studies have reinforced these findings. In 1998, a team led by occupational medicine expert Jens-Peter Bonde published a study of some four hundred first-pregnancy planners, examining the association between semen quality and the likelihood of conception per menstrual cycle. As total sperm counts increased up to 125 million, conception probability increased from zero to 25 percent, but no further increase occurred with higher sperm counts. Soon afterward, in 2002, a team led by public health researcher Rémy Slama examined the relationship between sperm counts and time to pregnancy in almost 1,000 fertile couples living in four European cities. Couples achieved pregnancy more and more rapidly up to a total sperm count of around 200 million, but not beyond that level. Complementing these findings, in 2010 andrologist Trevor Cooper and colleagues examined semen samples from almost 5,000 men in fourteen countries spread across four continents to provide reference values on behalf of the World Health Organization. For pregnancy to be achieved within a year, a lower limit of some 60 million sperms per ejaculate was identified. Overall, it seems that between 60 million and 200 million sperms per ejaculate are required for normal fertility. Although we still do not know why this is so, it is at least obvious that vast numbers of sperms are needed. Other mammals show similar requirements. An ingenious study of pregnancies arising from natural mating in sheep revealed a comparable fertility threshold of about 60 million sperms per ejaculate; below that threshold, the pregnancy rate dropped sharply from about 95 percent to around 30 percent.

How do sperms ever manage to reach the egg they need to fertilize? Again, it helps to look at our mammal relatives, as examination of their reproductive anatomy reveals an odd characteristic not present in humans—a bone inside the penis. The penis bone is unusual in two ways: It is by far the most variable bone between species, and it is never attached to any other bone. Most primates, bats, carnivores, insectivores, and rodents have a penis bone. It is particularly prominent in many carnivores. In a big dog, it can be some four inches long, but the acknowledged record holder among modern mammals is the walrus, which has a penis bone up to thirty inches in length. This awe-inspiring bone, known as an *oosik* in Alaska, has been widely used for carvings and rituals. Speaking of rituals, the originally all-male Tetrapods Club in London (to which I belonged long enough to vote for the admission of women) symbolically used a walrus penis bone as a gavel. On the other end of the spectrum, a raccoon penis bone is readily obtainable over the Internet, at least in the United States. Just type in "mountain man toothpick."

Not all mammal groups have a penis bone. Marsupials, rabbits, tree-shrews, elephants, sea cows, hoofed mammals, dolphins, and whales join humans in lacking this distinctive feature. Because of the penis bone's patchy distribution, its evolutionary history is uncertain. Perhaps the common ancestor of placental mammals had one and it was then lost from multiple lineages, or maybe it evolved independently in five or more groups of placental mammals. To add to the confusion, the penis bone is irregularly distributed among primates. Among lower primates, it is present and often quite large in all lemurs and lorises but absent from tarsiers. Most higher primates (simians) have a penis bone, but certain New World monkeys and humans are exceptions. All Old World monkeys and apes have one, although it is large in some species yet small in others.

A penis bone likely was present in ancestral primates and then retained, although somewhat downsized, in ancestral simians. A few Old World monkeys, such as mandrills and certain macaques, have a notably large penis bone, probably as a secondary development that harks back to the ancestral primate condition. The size of the bone differs markedly between macaque species, and this variation has been attributed to differences in mating behavior. As all apes have only a small penis bone, complete loss during human evolution ended a reduction process that was already under way. I once mentioned to an Australian colleague—an authority on penis bones in bats—that the complete absence of a penis bone in humans is an intriguing evolutionary puzzle. His laconic reply was: "Speak for yourself!"

Primatologist Alan Dixson has performed comparative studies revealing that a large penis bone is generally linked to prolonged mating. So it is reasonable to suggest that early primates, who had a large penis bone, mated for longer periods of time than modern primates, and as mating time became shorter, the penis bones became smaller, eventually disappearing completely in the human lineage. In other words, natural selection favored relatively brief copulation. Indeed, a large-scale multinational study published in 2005 by psychiatrist Marcel Waldinger and colleagues confirmed that human copulation lasts five minutes on average, although it may rarely last as long as forty-five minutes.

In addition to offering insight into the behavioral evolution of human copulation, the absence of a penis bone may also offer a new interpretation of the origins of the biblical story of Eve's creation. The tale—of Eve being created out of one of Adam's ribs—is of course figurative, but a 2001 paper by biologist Scott Gilbert and biblical scholar Ziony Zevit introduced a new angle. The Hebrew word *tzela*, translated as "rib" in the English version of the Bible, actually has several meanings, including a kind of supporting strut. Gilbert and Zevit proposed that the word *tzela* possibly referred to the penis bone, which is indeed absent, rather than to one of the ribs, which are all present and correct in a man's skeleton. Still, it is not clear how the originators of the biblical tale could have known that humans are unusual in lacking a penis bone. Among domestic beasts, all hoofed mammals also lack one. As already noted, it is prominent in dogs, yet quite small in cats. Thus any notion that the human penis has lost a bone must have come from comparisons with dogs.

EVERY MALE MAMMAL HAS a penis that can be stiffened for internal fertilization and, during mating, used to ejaculate millions of tiny sperms into the female reproductive tract. These two features of human reproduction date back to ancestral mammals more than 200 million years ago. Some basic features of sperm structure and production have even earlier origins. For instance, like other animals with backbones, mammals have a pair of testes containing sperm-producing tubules. Until released at ejaculation, sperms are stored in the tail end of a tightly coiled tubular annex, the epididymis. If laid out, a human epididymis would measure around twenty feet, the length of an average parking space, and sperms take two to three weeks to travel through it. Typically it holds a supply of about 400 million sperms.

Each tubule in the testis produces sperms in cyclical fashion. In humans the process takes about eleven weeks, the longest known duration for any mammal. Because individual batches of tubules are at different stages of sperm production at any one time, an active testis produces a continuous supply of mature sperms. In seasonally breeding species, the entire testis can be shut down and greatly reduced in size for part of the year; for example, most Madagascar lemurs have a strict breeding pattern and their testes shrink drastically outside the mating season. In the lesser mouse lemur, testis volume grows to ten times its normal size during the annual mating period. Other mammals, like humans, breed throughout the year, so the adult testis is continuously active. In the average human male, from sexual maturity onward, the two walnut-sized testes produce about 150 million sperms every day. This is equivalent to 1,500 sperms for every heartbeat and adds up to over 4 trillion over a typical male's life span. Nevertheless, although humans lack a restricted breeding season, there is still a detectable annual cycle in testosterone levels and sperm production, which I'll discuss in the next chapter.

A sperm always has three main parts: a head containing the nucleus, a midpiece packed with mitochondria (effectively a fuel tank), and a tail that propels the sperm for part of its journey toward the egg. Chromosomes are exceedingly tightly packed in the nucleus of the sperm head. Genes are inactivated because condensed, almost crystalline DNA is wrapped around special proteins. Although individual sperms are organized in the same basic way in all mammals, shape varies enormously between species. Astonishingly, although sperm size, too, varies quite widely between species, there is no general trend for an increase with body size: Sperms of mouse lemurs, rats, humans, elephants, and whales are all quite similarly sized. Indeed, there is actually a minor tendency for sperm size to decrease in larger-bodied mammals.

It is often claimed that the word "testify" comes from an ancient Roman custom in which a man would clutch his testes in his right hand before giving evidence in court. Regardless of the origins of this claim, it is undisputed that the Latin word *testis* originally meant "witness," and any human male can testify that the testicles are located in a risky place. To reduce the risk of injury, sumo wrestlers in Japan reportedly learn to massage their testes into the exits of the canals from the abdominal cavity, while British cricketers at bat wear a protective box to counter the threat of aggressively fast bowling with a hard missile. Descent of the testes into scrotal sacs outside the main body cavity is a truly extraordinary development that demands explanation. Why do human testes descend into such a hazardous position?

In all mammals, the reproductive and urinary systems develop in close association. As a result, the testes start to develop alongside the kidneys, high up in the belly cavity. Testes must therefore migrate backward and downward to end up in pouches outside the body. After moving through the belly cavity, each testis passes through an inguinal canal into an external scrotal sac. Carrying the testes outside the belly cavity is a special feature found in most mammals, although never in other animals. It is typical of all primates, including humans.

One explanation for descent of the testes, proposed in all seriousness, is that when higher energy turnover evolved in early mammals and they began to run around, their testes descended because of the pull of gravity. When I first read this suggestion, I at once pictured other, heavier body organs—heart, stomach, kidneys—swinging around in special pouches beneath the body, an absurd image. A related but somewhat more serious suggestion is that testes descend in some mammals to avoid concussion from pressures generated in the belly cavity by intense activity. Again, this reason would surely apply to other internal organs.

The generally favored explanation has been that testes descend to escape the raised core body temperature that characterizes all mammals. It is often stated that sperm production cannot occur at an elevated temperature. At first glance, some evidence appears to support this notion. When descent of the testes occasionally fails in humans, it is called cryptorchidism. Cryptorchid testes are generally well below normal size. About 3 percent of newborn boys have undescended testes, and in premature babies the condition is even more common. Although in 80 percent of cases the reluctant testes descend during the first year of life, and during later development in most of the remainder, if a testis remains in the belly cavity after puberty, sperm production is suppressed and surgical intervention is needed to restore fertility.

Human testes, like those of all other primates, reside outside the belly cavity at a lower temperature because of an adaptation that was already present in their common ancestor 80 million years ago. It is hardly surprising, then, that abnormal retention in the belly cavity suppresses sperm production, even if descent of the testes evolved for some other reason. As a counterpoint, there are many cases in which sperm production takes place despite an elevated body temperature. Birds, for example, never have descended testes, yet their average core body temperature is even higher than in mammals. Moreover, although most mammals have descended testes, there are many exceptions. In those animals, the testes remain in the belly cavity, often close to their original location near the kidneys.

Mammals that keep their testes in the abdomen do not have lower core body temperatures than mammals with descended testes, so we can dismiss the notion that an elevated body temperature must block sperm production. In fact, various findings indicate that testis descent is connected not with the production of sperms but with their storage. As already noted, mature sperms are stored until ejaculation in the tail of the epididymis, alongside the testis. In some mammals, the testis itself does not descend, but the tail of the epididymis migrates to end up against the ventral belly wall. Even in mammals in which both testis and epididymis descend, the epididymis leads the way and also consistently descends farther. Moreover, scrotal skin over the epididymis, but not over the testis itself, is often hairless and hence more easily cooled. It has been convincingly argued that the lower temperature in a descended epididymis increases availability of oxygen for stored sperms.

Evidence from birds also supports a link between sperm storage and the descent of testes in mammals. Birds store their sperms in seminal vesicles, functionally equivalent to the epididymis in mammals. In certain songbirds the seminal vesicles are contained in sacs close to the root of the penis, a location several degrees cooler than the core body temperature. Songbirds generally have relatively higher energy turnover than other birds, so perhaps sperms of some species are stored at a lower temperature for reasons similar to those that led to descent of testes in mammals. Nevertheless, the fact remains that most birds can both produce and store sperms at a relatively high core body temperature. So some special factor must have contributed to widespread descent of testes among mammals. Reliable storage may be a special challenge for mammal sperms because they need to travel a long way to fertilize a small egg.

The testis normally becomes active when a male reaches adulthood. Indeed, that is when testes usually descend into the scrotum in mammals. Primates are exceptional in this regard: Their testes are already descended at birth. In newborn males, the testes are at least low down in the belly cavity close to the inguinal canals, and typically have already migrated into the scrotum. In humans, the testes of a male fetus remain inside its abdomen until about the seventh month of pregnancy, but they are typically fully descended by the time of birth. Early descent of testes in primates is all the more surprising because, compared to other mammals, primates generally take longer to reach sexual maturity. Then why are primate testes descended at birth even though they will not produce sperms for some considerable time? The only available clue is that in all primates studied to date, male infants show a spike in the hormone testosterone around the time of birth. Perhaps this is an important new development in primate evolution, signaling an infant's sex and helping a mother to discriminate between newborn sons and daughters.

As mentioned above, it has long been recognized that heating harms human sperms. Indeed, Hippocrates mentioned this in two of his aphorisms. A man's testes in the scrotal pouch are only about 4°F cooler than his core body temperature, but that is enough to tip the balance between fertility and infertility. This raises the intriguing possibility that heating testes might reduce fertility and perhaps even furnish a simple method of birth control. Fertility expert John MacLeod, together with colleague Robert Hotchkiss, specifically tackled this topic in a 1941 paper. They applied dry heat using a fever therapy cabinet that enclosed a subject's whole body. Raising a man's body temperature while simultaneously increasing ambient temperature in this way led to a marked decline in sperm production. Disruption by heating took almost three weeks to take effect but then persisted for around two months.

It took some time before this technique was pursued further. Eventually, in 1965, fertility researchers John Rock and Derek Robinson published results from experimentally heating the scrotum in normal men. Among other things, they noted that immersion of subjects up to the neck in hot baths (110°F) made the scrotum about 2°F warmer than the core body temperature. In one experiment, for six weeks or more individuals wore insulating underwear that reduced the difference between scrotal and core body temperatures to 2°F instead of 4°F. In all cases, sperm counts began to decrease after about three weeks and reached their lowest levels between the fifth and ninth weeks. Low sperm counts persisted for three to eight weeks after the men stopped wearing the insulating underwear, and sperm counts returned to normal by three months after the treatment ended. Particularly interesting were the results obtained from the twenty men who started out with low sperm counts. Between two weeks and four months after the scrotum had been immersed in hot water for thirty minutes on six alternating days, sperm counts fell. In nine of the men, sperm counts later rebounded to higher levels than before treatment, and the wives of six of those men conceived within five months after treatment began.

A 1968 follow-up publication by Derek Robinson, John Rock, and Miriam Menkin reported on experiments in which the scrotum of human subjects was heated for thirty minutes by exposure to a 150-watt electric lightbulb on fourteen consecutive days. This procedure at first depressed sperm production, but temporary rebounds in sperm counts ensued. A couple of weeks after heating with the lightbulb, an ice bag was applied to the scrotum for about half an hour on fourteen consecutive days. This cold treatment, which reduced the temperature of the scrotum by just over 12°F, stimulated sperm production without any initial decline, nearly trebling the mean count measured at the outset. In sum, heating the testes decreased sperm counts, whereas cooling increased sperm production. These results suggest a direction for treating certain kinds of male infertility. Since Rock's time, a few studies have examined heating of the testes as a form of birth control. For instance, in 1992 surgeon Ahmed Shafik reported that wearing a polyester sling applied to the scrotum around the clock for an entire year suppressed sperm production after about five months. However, although the effect was shown to be reversible, the method has not led to any practical application.

HEATING THE SCROTUM, even by a few degrees, evidently can impair sperm production, so the next question to ask is whether certain activities or occupations might depress sperm counts. There have, for instance, been several medical reports that tight-fitting underwear can diminish semen quality. One study, reported by gynecologist Carolina Tiemessen and colleagues in the *Lancet* in 1995, specifically set out to test this. In a random sequence, nine volunteers—who all renounced hot baths, saunas, and electric blankets—wore either loose- or tight-fitting underwear around the clock for six months. Sperm counts and motility differed significantly between the two conditions. Average sperm counts were normal with loose-fitting underwear but almost halved with a tight fit. Sperm motility was affected even more, falling by two-thirds when men wore tight-fitting underwear.

Another suggestion that has been made is that driving a vehicle for extended periods may overheat the scrotum and reduce sperm counts. In 1979, fertility researchers Mihály Sas and János Szöllősi reported on a study of around 3,000 patients, including some three hundred professional drivers who showed substantially higher levels of disrupted sperm production. While deterioration of sperm production was relatively mild among car drivers, impairment was more pronounced in operators of industrial machinery and farm equipment. Moreover, the incidence of severe cases increased in proportion to the number of years spent driving. Among a hundred men who had been professional drivers for more than eight years, only four showed a normal sperm profile. Sas and Szöllősi discussed various reasons for disrupted sperm production, including pollution, but oddly did not raise the issue of scrotal heating.

Taxi drivers in Rome also became a focus of study. In a 1996 paper, a team led by industrial medicine specialist Irene Figà-Talamanca studied the effects of prolonged car driving. In comparison to controls, taxi drivers had a lower proportion of normal sperms (though sperm counts and motility were unaffected), and the effect was augmented with increasing time on the job. Subjects with poor sperms more frequently showed an extended delay before their partners became pregnant. Using statistical techniques, the researchers excluded various potential confounding factors. The good news is that one of those factors—moderate alcohol consumption—was actually associated with a small improvement in the semen profile.

Unsurprisingly, regular use of saunas can adversely affect sperm production. Appropriately enough, Finnish gynecologist Berndt-Johan Procopé published a pioneering paper on this topic in 1965. He observed a temporary, reversible decrease in sperm counts in a dozen men exposed to hot saunas for a total of about two and a half hours spread over a fortnight. Within a month or so after treatment ended, sperm counts had fallen by about half.

Men may overheat their testes in various other ways. A recent offender is a hot laptop computer, the effects of which were examined by urologist Yefim Sheynkin and colleagues in 2011. With twenty-nine healthy male volunteers, temperatures of the scrotum, of a laptop computer, and of a lap pad were recorded during three separate one-hour sessions, each with different conditions. Scrotal temperature increased under all three conditions, regardless of leg position or the use of a lap pad. The increase was lowest, only 2.5°F, when subjects sat with a pad beneath the laptop and their legs seventy degrees apart. The greatest increase, almost 4.5°F, occurred when the legs were held closely together without a lap pad, while an intermediate rise occurred when a lap pad was present. This study did not investigate the effects of scrotal temperature increases on semen quality, but studies of other kinds of heating indicate that prolonged use of a laptop may impair fertility, especially if the legs are pressed together and the computer is placed directly on the lap.

In some mammal species, social stress can disrupt normal testis function. An experimental study of treeshrews conducted by behavioral physiologist Dietrich von Holst revealed extreme effects. When two males are caged together, one becomes dominant over the other within a few hours and the subordinate male increasingly shows symptoms of stress. As exposure to the dominant continues, the effects become progressively more severe. One early response is retraction of the testes from the scrotum back into the belly cavity. If the dominant male is removed, the testes quickly return to the scrotum and no lasting harm is done. Longer-lasting testis retraction shuts down sperm production, and actual physical decay of the testes eventually follows—the submissive male is emasculated.

Responses to social stress are not so extreme in men. However, psychological stress may affect human testes and sperms more subtly, leading to reduced fertility. As is well known, infertility is itself a major cause of psychological stress, and so infertile couples may become locked in a vicious circle. Apart from that, various lines of evidence indicate that even moderate psychological stress can impair fertility in men, lowering testosterone levels and disrupting sperm production.

War theaters, of course, generate extreme psychological stress. One American study compared semen characteristics of more than three hundred Vietnam veterans with a similar sample of veterans who served elsewhere. Vietnam veterans had significantly lower average sperm concentrations and fewer sperms with a normal head shape, although sperm motility did not differ. Despite these observed differences in sperm features, Vietnam and non-Vietnam veterans reportedly fathered similar numbers of children.

In 2008, gynecologist Loulou Kobeissi and colleagues published another war-related example with more striking results. The authors used information from two infertility clinics in Beirut to investigate long-term impacts of the fifteen-year civil war in Lebanon. They compared 120 infertile men with 100 fertile men as controls. Infertile men were almost 60 percent more likely to have lived through the Lebanese civil war and to have experienced some kind of war-related trauma. The effect was attributed to wartime experiences and postwar exposure to various risk factors, including toxins and injuries, as well as stress.

Over the past four decades there has been an animated debate about the alarming possibility that human sperm counts have declined markedly since the 1950s. In 1974 fertility experts Kinloch Nelson and Raymond Bunge examined semen specimens from almost four hundred men who had chosen to undergo vasectomy (severing of the sperm-carrying ducts) between 1968 and 1972. The average sperm count per ejaculate before vasectomy was found to be about 135 million sperms. Nelson and Bunge noted that this was much lower than counts recorded by earlier investigators, which averaged over 300 million. Because of the startling difference between these earlier studies and their own, Nelson and Bunge reviewed semen analyses from four hundred men who had had an infertility evaluation done at their hospital between 1956 and 1958. A quarter of those men had total sperm counts above 300 million. After carefully excluding various alternative possibilities, Nelson and Bunge concluded that "something had altered the fertile male population to depress the semen analysis remarkably."

This claim, along with similar reports, rapidly led to controversy. Scientists on both sides of the debate based arguments on data from a few large samples examined at different times and often in different places. Eventually biologist William James, who has conducted careful statistical studies of human reproduction over the past fifty years, stepped in to test whether there really was a trend in reported sperm counts. He tracked down representative data on average sperm counts of unselected men over a period of forty-five years and presented results in a 1980 paper. His conclusions were unequivocal: "There can be no reasonable doubt that the reported mean sperm counts show a decline with time of publication, at least since 1960."

Some critics have suggested that changes in methods of semen analysis over the years could have affected human sperm counts, and this possibility deserves serious consideration. But Australian veterinarian Brian Setchell elegantly eliminated this potential problem. Comparable methods have been widely used to determine sperm numbers in the semen of farm animals, so they should also show any decline due to changing techniques. Sperm counts dating back to the early 1930s are available for cattle, pigs, and sheep. In a 1997 paper, Setchell reviewed data from over three hundred publications reporting sperm counts over the period from 1932 to 1995. There was no significant change in sperm counts over time for either bulls or boars, but with sheep there was actually a slight, but significant, increase. As Setchell sagely concluded: "It would appear that, if the fall in human sperm counts is real, then it must be due to something which is not affecting farm animals."

Reports of declining human sperm counts have continued to accumulate around the world. Two recent reports have provided convincing evidence of marked declines in sperm counts over the past twenty years in Israel and France. Entirely new information for Israel was published in a 2012 paper by a team led by Ronit Haimov-Kochman, an Israeli infertility researcher. They reported results of a retrospective analysis of over 2,000 weekly sperm samples collected from fifty-eight young paid donors over the fifteen-year period from 1995 to 2009. Average sperm counts dropped significantly by almost 40 percent, from over 300 million sperms to around 200 million. As a result, it has become increasingly difficult to find sperm donors who meet the criteria set by fertility clinics. Haimov-Kochman and her colleagues concluded that this rapid deterioration of semen quality among fertile semen donors may shut down sperm donation programs. The second 2012 paper, published by epidemiologist Joëlle Le Moal and colleagues, identified a similar decline in semen concentration through a countrywide survey in France. This study, which was also retrospective, examined sperm counts for almost 27,000 men participating in assisted reproductive technology procedures because their partners were totally infertile. Over the seventeen-year period from 1989 to 2005 there was a continuous decrease in semen concentration of almost 2 percent per year. Across the study period the overall decline was some 32 percent, falling from about 220 million sperms per ejaculate in 1989 to under 150 million in 2005.

Even more alarmingly, decreasing sperm counts seem to have been paralleled by increasingly frequent abnormalities of the male reproductive system, including cryptorchidism, penis malformation, and testicular cancer. This trend raises the possibility that the factors responsible for decreasing sperm counts may be having increasingly serious impacts on the male sex organs themselves. As testimony to a possible link, a research team in Denmark noted that in Danish men the incidence of testicular cancer is five times higher than in Finnish men, while their sperm counts are more than 40 percent lower. The striking decline in semen quality and increase in abnormalities of the male sex organs detected over only fifty years are surely due to environmental factors, not to genetic change.

One reason for controversy about declining sperm counts is that reported trends differ from population to population. Even countries that are not very far apart, such as Denmark and Finland, can differ greatly. This suggests that an environmental influence such as pollution may be responsible, and some evidence does indeed link declining sperm counts to local toxins. In 2006, obstetrician Rebecca Sokol published a striking study conducted in Los Angeles, home of the saying "Any air we can't see we don't trust." Seeking a link to environmental pollution, the Sokol group analyzed sperm bank semen samples collected over a three-year period from forty-eight repeat donors. Analysis revealed one consistent significant finding: As ozone levels rose, sperm concentrations declined. If the yawning man-made ozone hole in our atmosphere has not sufficiently grabbed your attention before now, this might just do the trick.

In fact, there are several candidates for environmental factors that might reduce sperm counts. For instance, some evidence indicates that smoking and immoderate drinking during pregnancy impairs sperm production in sons later in life. It has only recently been recognized that chemicals in widely used synthetic products are also strong candidates for sperm-suppressing toxins. An increasingly prominent example is bisphenol A (BPA), an organic ingredient used to make polycarbonate plastics, epoxy resins, and various other everyday items such as DVDs, sunglasses, medical devices, automobile parts, sports equipment, and glazing. Heat-resistant polycarbonate plastics are widely used to package food and drink as well as to coat the insides of cans. BPA ranks among the top fifty chemicals now produced, with global annual production exceeding 5 million tons by 2008. Yet it has only recently come under public scrutiny. A 2010 report from the

U.S. Food and Drug Administration raised concerns about exposure of fetuses, infants, and young children, and Canada became the first country to recognize BPA officially as a toxic substance. In parallel, alarming reports of the noxious effects of BPA have accumulated rapidly. But several authorities have questioned the evidence that BPA has toxic effects. The official stand in the European Union remains that food contact is not a hazard, although a new risk assessment was launched in 2012. One thing is certain, however: Everyone in industrialized nations is exposed to BPA every day, as the substance leaches from containers into food, especially when heated, and it is universally detected in blood and urine. It is particularly alarming that, compared to adult levels, exposure to BPA is ten times greater for babies in neonatal intensive care units and doubled for children generally. In one notable experiment, when seventy-seven Harvard student volunteers drank cold liquids exclusively from baby bottles for just one week, levels of BPA in their urine samples rose by more than two-thirds.

In 2010, a Chinese-American team led by reproductive epidemiologist De-Kun Li reported on a study of self-reported sexual problems in men exposed to BPA at the workplace. The team interviewed subjects employed in factories in China, some exposed to exceedingly high levels of BPA, others unexposed. After carefully excluding effects of several possible confounding factors, Li and colleagues found that the risk of problems was consistently four to seven times higher in workers exposed to BPA than in controls. Deficiencies spanned every domain of male sexuality: Sexual desire, erection, ejaculation, and sexual satisfaction were all impaired. Moreover, the risk of sexual problems increased with levels of cumulative exposure to BPA. Workers in factories with exposure to BPA also reported higher frequencies of sexual deficiencies within a year of employment. It should be emphasized that factory exposure to BPA was extreme; levels in urine samples from workers were some fifty times higher than in controls.

A year later, members of the same research team reported results from a follow-up study that directly examined semen quality in relation to urinary BPA levels. More than two hundred men with and without workplace exposure to BPA were studied in four regions of China. An increasing BPA level in the urine was significantly associated with decreased semen quality even after allowing for confounding effects. In subjects with detectable urinary levels of BPA the risk was more than doubled for lower sperm motility, more than tripled for decreased sperm concentration and vitality, and more than quadrupled for lower sperm counts. These results provided the first evidence that BPA adversely affects semen quality.

While attention has mainly focused on BPA ingested through food or drink, this chemical can also be absorbed directly through the skin. Therein lies the rub. BPA is often used on paper receipts printed with lightweight devices that use a thermal transfer process. This process has been widely used in cash registers and ATMs since the 1970s. As a result, BPA has also become a major contaminant of recycled paper. Thermal paper does not always carry BPA, but a powdery layer of this chemical is often used to coat one side of a receipt. In 2010, large-scale laboratory tests commissioned by the Environmental Working Group in the United States found high levels of BPA on four out of ten receipts sampled from major businesses and services. Total amounts of BPA detected on receipts were up to a thousand times greater than other, more widely discussed sources of exposure, such as plastic bottles and food cans. Laboratory wipe tests showed that BPA is easily removed from receipts, so it can surely rub off on the hands of anyone handling them. Millions of salespeople work at cash registers, and every one of them may handle hundreds of BPA-coated receipts in a single day. Monitoring by the federal Centers for Disease Control revealed that salespeople carried an average of 30 percent more BPA in their bodies than other adults.

BPA has been with us for quite some time, as it was first synthesized in 1891. Yet alarm bells should have been ringing since the 1930s, when BPA experimentally applied to female rats that had had their ovaries surgically removed was found to behave like a steroid hormone. When BPA (or any of a dozen similar compounds) was fed to female rats without ovaries, the lining of the vagina was affected in the same way as with estrogens. Thus we have known for more than seventy years that BPA acts like an estrogen. Yet authorities argue that BPA is less potent than real estrogens and is rapidly broken down and eliminated from the human body. For these reasons, they say, it is unlikely to pose a significant risk to our health. The website of the Polycarbonate/BPA Global Group of the American Chemistry Council indicates that BPA is entirely safe. In 2008, the U.S. Food and Drug Administration echoed this appraisal, but Congress sensibly took steps to restrict the use of BPA and asked the FDA to reexamine the issue. As yet, no definitive action has been taken. However, we sorely need to know how environmental toxins such as BPA affect not just sperms but also eggs and the ovaries that produce them. As a first step in that direction, we must understand how the female sex organs and cells evolved. So far this chapter has focused on male aspects of reproduction, but that is just a prelude to the far greater contribution made by all female mammals.

As is the case with testes and sperms, we share basic features of human ovaries and eggs with all mammals. On each side of the female body, there is a single ovary. Like the testis, it develops in close association with the adjacent kidney, but it is smaller. A human ovary is roughly the size of an almond, about a third of the volume of the testis. Moreover, ovaries never shift far from their starting position in the belly cavity.

Each ovary lies close to the funnel of a tubular oviduct called the Fallopian tube, leading to the womb. Once an egg has been released from the ovary, in a process known as ovulation, it enters the oviduct funnel and begins to migrate down toward the womb. It first encounters a part of the oviduct known as the ampulla, which has a richly folded inner lining. To reach the womb, it must then pass through the next section of the oviduct, the isthmus, which has a smoother lining. Fertilization of the egg by a single sperm typically takes place close to the junction between the ampulla and the isthmus.

In most mammals, each ovary is enveloped by a pocket of tissue, the ovarian bursa. This enclosure of the ovary—with the exception of a small opening to the belly cavity—ensures that the egg safely enters the oviduct funnel to start its journey to the womb. As lemurs, lorises, and most nonprimate mammals have an ovarian bursa, this is most likely the primitive condition. But in tarsiers, monkeys, apes, and humans, the bursa has vanished. At first sight this is puzzling. The loss of the bursa during evolution would be expected to increase the risk that an egg will escape into the belly cavity instead of safely entering the oviduct.

On rare occasions, an egg indeed fails to enter the oviduct in humans. If the egg is nonetheless fertilized, the embryo develops in the belly cavity. Attachment of an embryo in the wrong place, which happens in about one in every hundred cases in humans, is called an ectopic pregnancy. In most cases, the embryo develops in the oviduct. Failure of the egg to enter the oviduct, leading to the development of the embryo in the belly cavity, is much less frequent, occurring in only 1 in 10,000 human pregnancies. Of course, without surgical intervention, this condition is fatal for both mother and baby. Strong selection should favor efficient transfer of eggs from the ovary to the oviduct in all mammals, so why on earth did the ovarian bursa disappear in the ancestor of tarsiers, monkeys, apes, and humans? Its loss must have been accompanied by development of some special mechanism that replaced the original enclosing function.

Indeed, there is good evidence of such a replacement mechanism. Ovulation in rhesus monkeys and women has been directly observed using a laparoscope, an instrument that permits a physician to peer inside the belly cavity. Such examination revealed that the oviduct funnel maintains close contact with the ovary and actively moves across its surface, groping around to seek the site where ovulation will occur. This movement of the oviduct funnel may be what ensures efficient and timely transfer of the egg to the oviduct. As all members of this group of primates lack an ovarian bursa, it is reasonable to infer that the sweeping motion of the oviduct funnel is also universal within the group. Still, this mechanism makes sense only if ovulation from each ovary is limited to a single egg at a time. Without an ovarian bursa, multiple ovulations from a single ovary would magnify the risk of an egg going astray. Thus the common ancestor of monkeys, apes, and humans was presumably adapted to produce only one or two infants at a time. As we shall see in later chapters, other lines of evidence confirm this inference.

Also connected to the limiting mechanisms of childbearing is the fact that the maximum number of eggs that any woman can produce is limited from the outset. Eggs generally develop in waves from starter cells known as oogonia. With few exceptions, a female mammal starts out with a basic stock of starter cells that is gradually depleted over her lifetime. In the human female, each ovary attains its peak number of starter cells—about 7 million—halfway through fetal development. By the time of birth, the supply has already decreased to around 2 million, and in a seven-year-old girl only 300,000 starter cells are left. Of those, only a few hundred will go on to release mature eggs. Consequently, the number of sperms in a single ejaculate is half a million times greater than the maximum number of eggs that a woman's ovaries can produce during her reproductive years.

Whereas individual sectors of the testis show different stages of sperm production, egg production in mammals typically follows a single cycle in which the two ovaries act in concert. Starting from an oogonium, each egg cell, or oocyte, develops within the ovary inside a cluster of cells known as a follicle. As the egg cell ripens, the follicle grows in size and a fluid-filled cavity eventually forms inside. At this stage, the enlarged follicle migrates to the surface of the ovary, where it can release its egg. Maturation of a follicle can be arrested at any stage, in which case the follicle degenerates, a process known as atresia. Even in mammals with single ovulations, including women, several follicles usually enter the final maturation phase in both ovaries in each cycle. Development of each follicle takes more than a year, almost four hundred days. New follicles begin to develop continuously, so at any given time all stages are present in the ovary. In each human cycle, a clutch of follicles is recruited to begin the final phase of maturation, but through an unknown process a single follicle in one of the two ovaries eventually becomes dominant in most cases. This dominant follicle usually proceeds to ovulation, while all others degenerate. In some cycles, however, not a single follicle makes it as far as ovulation.

In the ovarian cycle of any mammal, development of follicles is driven by follicle-stimulating hormone (FSH), produced by the pea-sized pituitary gland, located beneath the brain. As they develop, the follicles themselves produce steroid hormones, notably estrogens. Release of the egg from a ripe follicle at ovulation is typically triggered by a marked spike in luteinizing hormone (LH), another hormone produced by the pituitary gland. Following ovulation, the empty husk of the follicle is converted into a yellow body (corpus luteum), and moderate levels of LH stimulate it to produce progesterone. If conception does not ensue, the corpus luteum persists for a limited period and then degenerates before the next cycle begins. Hence an ovarian cycle can be divided into an initial follicular phase, in which follicles develop up to the point of ovulation, and a subsequent luteal phase, with formation of a yellow body after ovulation.

In a woman, the transition from the follicular phase to the luteal phase is marked by a small but detectable upward shift in her basal body temperature, which is the minimal level found while she is resting. In fact, this rise of 0.5°F to 1°F in basal body temperature in the middle of the cycle reflects an increase in energy turnover that is then maintained for the rest of the cycle. Because the temperature rise commonly occurs soon after ovulation, it has often been used as an easily measured indicator of the release of an egg from the ovary. Of course, sensitive hormone assays now provide more reliable information, but basal body temperature is still used as a rough-and-ready guide to ovulation.

* * *

Although the basic structure of the ovarian cycle is the same in all mammals, there is a key distinction in the relationship between mating and ovulation. In some mammals, such as cats, rabbits, and treeshrews, the act of mating itself triggers ovulation by provoking a surge in LH. Biologists call this induced ovulation. If the female does not mate during a cycle, there will only be a follicular phase; ovulation will not occur and no yellow body will be formed. These cycles without mating are short. In a minor variation, in some mammals (such as mice) ovulation occurs without mating, but mating is needed for a yellow body to form. The outcome is the same in both cases: A yellow body is formed only if mating has occurred. For simplicity, we can lump the two cases together as "induced ovulation."

Induced ovulation stands in direct contrast with the condition in other mammals, such as humans, that ovulate regardless of mating. In each cycle, an LH surge is generated internally, triggering ovulation followed by formation of a yellow body. Mating is not needed to stimulate ovulation, which is said to be "spontaneous." A cycle that does not depend on mating for ovulation or formation of a yellow body is typically long because it automatically comprises both follicular and luteal phases. Thus we can distinguish between mammals with short, mating-dependent cycles and mammals with long cycles that are independent of mating.

Members of each major group of mammals tend to have a particular type of ovarian cycle. For instance, carnivores, insectivores, rodents, and treeshrews generally have short, mating-dependent cycles, whereas primates, hoofed mammals, dolphins, and elephants typically have long, mating-independent cycles. Women, like all other primates, have a cycle in which spontaneous ovulation is followed by automatic formation of a yellow body, so this feature was likely already present in their common ancestor more than 80 million years ago. Average cycle length across primate species is about a month overall, as in women, so the duration of the ovarian cycle in ancestral primates was probably close to this length.

In 1554, at the age of thirty-eight, Queen Mary I of England married Philip II of Spain. Given her age and situation, she was more than eager to have a son. Just two months after the wedding one of her physicians announced a royal pregnancy, reportedly accompanied by outward signs such as gradual swelling of her belly and morning sickness. Nine months later, however, there was nothing to show the world. It has been suggested that acute disappointment at this shocking outcome triggered the persecutions that earned the queen the nickname "Bloody Mary." After three years had elapsed, the same thing began all over again. This time Queen Mary's health declined, contributing to her early death.

It is widely believed that Queen Mary experienced phantom pregnancy, also known as hysterical pregnancy, a rare condition reflecting a serious emotional and psychological disorder. Sigmund Freud's memoirs report that his best-known patient, "Anna O," falsely believed that she was pregnant with the child of her previous psychoanalyst, Josef Breuer. Although there is much variation in symptoms, sufferers from phantom pregnancy generally have one thing in common: an extreme longing to bear a child. Most cases occur in women in their thirties or forties. Phantom pregnancy is extremely rare in women, currently accounting for 1 in 7,000 pregnancies in the United States. Many natural signs of pregnancy, such as cessation of menstruation, mood swings, cravings, morning sickness, enlarged and tender breasts, abdominal distension, movement within the womb, and weight gain, may be present with a phantom pregnancy.

In mammals with induced ovulation, phantom pregnancy expresses itself quite differently. If mating leads to fertilization, as generally occurs under natural conditions, progesterone produced by the yellow body helps to establish and maintain the ensuing pregnancy. If mating is unsuccessful, production of progesterone by the yellow body serves no useful purpose, commonly leading to a phantom pregnancy. This often happens in cats when a female is mated by an infertile male. During phantom pregnancy in such mammals, a female undergoes some of the changes accompanying real pregnancy, such as boosting of the blood supply to the womb and thickening of its inner lining. At some stage, though, the female's body recognizes that no developing offspring are present and the phantom pregnancy comes to an end. When this happens, blood and debris are often shed from the womb, as occurs in menstruation. Yet, as I will explain below, there is a fundamental distinction between shedding of tissue following a phantom pregnancy and the special case of menstruation.

Over the course of four to five days at the end of her ovarian cycle the average woman sheds about a fluid ounce of blood. This amounts to a loss of almost a pint of blood—and for some women four times that—in a year. Bleeding results from shedding of the inner lining of the womb as the yellow body fades and the luteal phase winds down. It became known as "menstruation," from the Latin *mens* for "month," because it occurs at approximately monthly intervals. Customarily, the word has been used to refer to species with external bleeding, although some authors misleadingly refer to ovarian cycles of all primates as menstrual cycles. While primates across the board typically have long ovarian cycles, there is a major distinction regarding blood loss from the womb at the end of the cycle. Lemurs and lorises lack such bleeding because the placenta does not invade the womb's inner lining; they do not even experience shedding from the womb at the end of pregnancy. Menstrual bleeding is limited to monkeys, apes, and humans, and women have by far the most pronounced menstruation among these primates.

Menstrual cycles have been extensively studied in Old World monkeys, apes, and humans, but it is not clear why blood loss from the womb occurs at all. In contrast to lemurs and lorises, the placenta of tarsiers, monkeys, apes, and humans is highly invasive and establishes direct contact with maternal blood in the wall of the womb. Various changes take place in the womb's inner lining during the luteal phase of the ovarian cycle. The changes, including the rapid sprouting and marked enlargement of blood vessels, prepare the womb for attachment and initial development of the placenta if conception occurs. The womb's inner lining, known as the endometrium, also becomes markedly thicker during the luteal phase. If conception does not occur, the luteal phase ends in menstruation. For years menstruation in humans, apes, and monkeys was commonly seen as nothing more than shedding of redundant tissue from the womb.

In 1993, Margie Profet proposed a radical explanation for the evolution of menstruation that attracted considerable media attention. In a widely cited paper, she proposed that menstrual bleeding protects the womb and oviducts against germs carried by sperms. Her theory was appealing, as pathogens hitchhiking on sperms or in the seminal fluid would, of course, pose a serious threat to the female reproductive tract. Yet, as further consideration revealed, Profet's idea was fundamentally flawed. Problems caused by sperm-borne germs would be expected to be virtually universal among mammals, in which case menstruation should also be universal. Profet claimed that universal menstruation did occur, but this claim is simply false. Apart from monkeys, apes, and humans, true menstruation has been securely demonstrated only for certain bats and odd little African mammals called elephant shrews.

Furthermore, natural selection would surely favor countermeasures to protect the womb against germs riding on sperms. Indeed, it seems that one function of the abundant mucus secreted from the neck of the womb in mammals is to block invasion by germs deposited in the vagina during copulation. Moreover, white cells are commonly present in the vagina to counter invading germs. In sum, no convincing evidence connects avoidance of sperm-borne germs with the evolution of true menstruation. Indeed, this hardly seems a logical possibility because human copulation can occur up to four weeks before the next menstruation. With such a long interval between copulation and menstruation, germs would have plenty of time to multiply and spread throughout the womb and oviducts.

At present, we are left without a widely accepted explanation for the evolution of menstruation. Any suggested interpretation must account for the heavy blood loss in women. A fluid ounce of blood loss with each menstrual flow translates into an appreciable quantity of iron, an essential and often scarce mineral. Particularly heavy menstrual bleeding can lead to anemia or intensify preexisting iron deficiency. One investigation of women in Brazil showed that body stores of iron become depleted in women who shed more than 2 fluid ounces of blood per cycle, and clinically recognizable anemia is likely when menstrual blood loss rises above 3 fluid ounces. Clearly, a substantial selection pressure must have favored the origin and maintenance of such heavy bleeding during human evolution, contrasting starkly with the lesser blood loss shown by apes and monkeys.

Anthropologist Beverly Strassmann has proposed that menstruation evolved as an energy-saving adaptation. Cyclical thickening and thinning of the womb's inner lining is universal among mammals, but it is pronounced in monkeys, apes, and humans. Energy consumption of the womb's inner lining has been measured in humans by monitoring tissue slices, and it rises to seven times the starting level by the end of the luteal phase. In Strassmann's view it is more costly to sustain the expanded womb lining than to regenerate it in each cycle. She suggested that bleeding is a side effect that arises when the amount of blood is too great for efficient absorption. She also noted an additional flaw in Profet's hypothesis: In societies without contraception, menstruation is a rare event among women during their reproductive years. Strassmann's perspective on human menstruation is enriched by her long-term fieldwork with the Dogon in Mali. An average Dogon woman has about nine pregnancies leading to live births, and menstruation is relatively rare, especially during the prime childbearing years. Even so, Strassmann's energy-economy hypothesis does not explain why human blood loss during menstruation is so much heavier than in any other primate.

In 2009, a group led by gynecologist Jan Brosens proposed an ingenious new explanation for menstruation. Recognizing that true menstruation is restricted to monkeys, apes, humans, and a few other mammals, the researchers noted that, like pregnancy, it is an inflammatory condition. Accordingly, they suggested that menstruation serves to precondition the uterus to respond to the deep invasion of maternal tissues that occurs during pregnancy. One strength of this proposal is that the heavy loss of blood during human menstruation can be linked to the especially invasive nature of human pregnancy. This explanation may help to demystify some of the disorders of human menstruation and pregnancy.

One final possibility merits consideration, although evidence supporting it is limited. In mammals that have true menstruation, sperms may be stored somewhere in the female reproductive tract. It follows that the shedding of blood and debris from the womb might serve to flush out timeworn sperms at the end of an ovarian cycle.

It is generally accepted that sperm survival is limited, with a typical upper limit of two days in most mammals, including humans. Yet it has been known for some time that in humans large numbers of sperms are stored in special small pouches or crypts in the neck of the womb. These crypts also play a key part in production of mucus, a watery gel that has figured prominently in discussions of the human cycle and fertility. Around midcycle, the mucus has a thinned consistency similar to egg white, and this has been taken as an approximate indicator of ovulation time.

Biophysicist Erik Odeblad has identified four basic types of human mucus (G, L, P, S), each with a different function. Different mucus types are produced by particular crypts in various regions of the neck of the womb, and their proportions vary across the menstrual cycle. During the buildup to ovulation, L mucus blocks the passage of abnormal sperms, while S mucus guides normally formed sperms into crypts in the neck of the womb. Once in the crypts, sperms are temporarily sealed in with mucus plugs. Around the time of ovulation, P mucus dissolves the plugs. Sperms are released from the cervical crypts and can move on toward the oviduct. Early in the cycle and during the luteal phase following ovulation, G mucus forms a barrier at the lower end of the neck of the womb that sperms cannot easily penetrate.

In short, human sperms may survive intact for several days in mucus produced by the neck of the womb. Reproductive biologist John Gould showed that sperms were found to retain fertilizing capacity up to eighty hours after insemination. Indeed, motile sperms with normal swimming speeds have been recovered up to five days after insemination. In another study, a team led by gynecologist Michael Zinaman examined sperms in mucus collected from women up to three days after artificial insemination. In all cases, most of the recovered sperms were found to be viable. They concluded that sperm function is conserved in the neck of the womb, indicating that it is a sperm storage site in women.

A lot is known about mucus produced by the neck of the human womb, and a little about sperms found in mucus. Regrettably, we know remarkably little about the actual storage of sperms in the associated crypts. Our only knowledge comes from a landmark 1980 study by gynecologist Vaclav Insler and colleagues in which twenty-five women scheduled for surgical removal of the womb (hysterectomy) volunteered to be artificially inseminated on the day before their surgery. They were split into three groups: Nine women were pretreated with estrogen and inseminated with normal semen, nine were pretreated with a progesterone-like hormone and inseminated with normal semen, and seven were pretreated with estrogen and inseminated with abnormal semen. After the womb was removed on the day after insemination, sperm storage in the crypts was examined microscopically in serial sections of the neck of the womb.

Crypts varied greatly in size, and sperms were mainly stored in the larger ones along the neck of the womb. Indeed, within two hours after insemination, sperms colonized the entire length of the neck. Insler and his team calculated the number of crypts containing sperms and sperm density per crypt. Both the proportion of colonized crypts and sperm density were significantly higher in wombs that had been pretreated with estrogen than in those pretreated with the progesterone-like hormone. With estrogen treatment, up to 200,000 sperms were found stored in the crypts of a single womb, whereas following treatment with the progesterone-like hormone the maximum was only a quarter of that number. They also discovered that semen quality was critically important for sperm storage. Both the percentage of crypts containing sperms and sperm density were severely reduced in the wombs of volunteers inseminated with abnormal semen. In a nutshell, this evidence indicates that sperm storage in crypts in the human womb is more likely during the follicular phase, when estrogen predominates, and that healthy sperms are more likely to be stored than abnormal sperms. In fact, Insler and colleagues reported that live sperms have been found in cervical mucus up to the ninth day after insemination. Taking the evidence as a whole, they suggested that the neck of the womb serves as a sperm reservoir from which viable sperms are gradually released for migration into the oviduct. Such slow, controlled release could ensure the survival of sperms capable of fertilization over an extended period.

There is also some evidence indicating that sperm may be stored for a while in the oviduct as well as in the neck of the womb. It is known that sperms can attach to the lining of the oviduct in humans and various other mammals. Reproductive physiologist Joanna Ellington and her colleagues studied such attachment of sperms to surface cells isolated from the oviduct. They found that unattached sperms showed various abnormalities compared to sperms that remained unattached. So attachment to the oviduct lining might provide an additional basis for selection of higher-quality sperms.

AT THIS POINT, we need to make sense of the cycles in the human ovary. It is standard medical practice to take the beginning of menstrual bleeding as day one of a cycle and to define the end of the cycle as the last day before the next menstruation. Research papers and medical textbooks commonly refer to an idealized cycle lasting about four weeks, with ovulation occurring midway between two menstruations. Regular monthly cycles with midcycle ovulation are seen as the norm; marked deviations are treated as abnormalities. However, this standardized "clockwork egg timer" model is an abstraction and can be misleading. In real life, there is wide variation, which is biologically significant in its own right.

A classic 1967 paper by reproductive biologist Alan Treloar and colleagues assessed variation in menstruation across the human life span. During the first five years after menstrual bleeding begins, at menarche, cycle length is quite variable and the average is greater than during the years of peak fertility. Cycles then settle down to a more regular pattern, and their length gradually decreases with age for about twenty-five years. After that, cycle length increases again and becomes substantially more variable toward the end of reproductive life, during the transition to menopause. By that point, average cycle length has increased to eight weeks instead of the previous four. During the main fertile years of twenty to forty-five, overall yearly averages for the menstrual cycle do stay quite close to four weeks. On the other hand, individual yearly averages vary considerably between women, with an overall range from twenty-six to thirty-one days. Not unexpectedly, variation from cycle to cycle in individual women is even greater. A spectrum from three to five weeks is typical, but occasional cycles fall outside this range.

In another study, gynecologist Kirstine Münster and colleagues analyzed cycle variation in an entire county. In Denmark, school health education promotes the keeping of menstrual calendars, and women in the survey, ages fifteen to forty-four, were asked to submit their records for 1988. With few exceptions, the usual cycle length reported by individual women ranged between twenty-one and thirty-five days. However, in a third of women cycle length varied by more than fourteen days during the year. Interestingly, cycles were more variable in women from lower socioeconomic classes, indicating that environmental factors can magnify baseline variation. In medical circles, a ten-day range—from twenty-three to thirty-three days—is often taken as standard, and many gynecologists see any greater variation as pathological. As Münster and colleagues concluded: "If this were true, about two-thirds of normal Danish women . . . would be suspected of disease or at least some kind of . . . disturbance." We must bear in mind that the human menstrual cycle varies markedly, and we should not be surprised when women do not adhere to the clockwork egg timer model.

As with sperm production, psychological stress can affect ovarian cycles. Here, too, Dietrich von Holst's studies of social stress in treeshrews are instructive. If two adult females are kept together in a cage, one of them becomes dominant within a few hours. The subordinate female then shows changes that gradually increase in severity as exposure to social stress from the dominant continues. Within a few days, ovarian cycles cease and the ovary shuts down completely. There are other primates that exhibit similar changes. In the common marmoset, a small New World monkey, suppression of ovulation occurs in subordinate females in a social group, although the effects of subordinate status are otherwise less severe. In both treeshrews and marmosets, subordinate females stop ovulating as an adaptation to conserve resources, hedging their bets by deferring breeding until social conditions improve. It is a safer option to suspend ovarian activity than to become pregnant and suffer the consequences of aggression from the dominant female.

A connection between psychological stress and abnormal menstrual cycles has long been recognized in women. Circumstantial evidence of various kinds indicates that certain conditions can adversely affect ovarian activity. A stark example emerged in the throes of World War II, when hundreds of British women were interned in a Japanese concentration camp in Hong Kong in 1941. Gynecologist Annie Sydenham, herself an internee, later reported that half of the women between the ages of fifteen and forty-five showed complete arrest of menstruation. This change occurred as soon as the women were interned, well before malnutrition could have had any impact, and persisted for three to eighteen months. Sydenham reasonably attributed menstrual arrest to the emotional shock of war and internment.

Another convincing example of the impact of severe stress on the menstrual cycle comes from a 2007 study of women beginning lengthy jail sentences. Epidemiologist Jenifer Allsworth studied almost 450 women of reproductive age. Disruption of menstrual cycles was common in this prison population: Almost a third of women reported some kind of irregularity, and one in ten completely failed to menstruate for three months or more. Many of these women had, of course, experienced trauma long before starting their jail sentences. More than half had suffered some kind of sexual abuse, either in childhood or as adults. Factors such as having a parent with a history of alcohol or drug problems and exposure to physical or sexual abuse almost doubled the likelihood of experiencing menstrual irregularities in jail. Allsworth also found a significant difference regarding previous pregnancies between women with menstrual disruption and those with regular cycles. Although the numbers of pregnancies were similar in the two groups of women, notably fewer women with irregular menstrual cycles carried pregnancies to term.

It is now widely accepted that exposure to severe psychosocial stress can disrupt menstrual cycles. But other factors such as physical exercise, weight loss, and diet are also known to influence ovarian activity, making it difficult to establish a convincing link between moderate psychological stress and more subtle menstrual irregularities. Moreover, it is also possible that ovarian cycles vary according to time of year, adding a complicating factor to the interpretation of menstrual cycles in women. Ultimately, we must understand seasonality if we are to decipher how the menstrual cycle evolved, and gain a deeper understanding of the mechanics of conception and the human reproductive cycle.

CHAPTER 2

Cycles and Seasons

Evolution takes place in natural habitats, so that is where we must look to see how features of behavior and physiology, including seasonality and cycles, emerged in any species. In 1968, I had my first opportunity to observe primate breeding behavior in the wild, which allowed me to witness seasonal patterns firsthand. My earlier research had revealed that lesser mouse lemurs, little creatures weighing just two ounces, share many features of the common ancestor that gave rise to all primates, including humans. After some basic observations of captive mouse lemurs in a breeding colony that I had established at University College London, I traveled to Madagascar to study these tiny primates under natural conditions.

I was eventually able to piece together an overall picture of the natural breeding system. Adult females become receptive in rapid succession in late September and early October. Most conceive at once and give birth in late November or early December, after a pregnancy lasting about two months. As the mating season approaches, the testes of males expand, reaching ten times their resting size by the time they begin copulating. Confirming reports from several previous fieldworkers, I found that timing of the breeding season of lesser mouse lemurs is consistent from year to year at any given site. Indeed, when I returned to Madagascar in 1970 to do more work at the study area, breeding followed exactly the same pattern.

This consistency raises the obvious questions: Why do births occur at a particular time every year, and what prompts males and females to mate at the appropriate time? A clue lies in the fact that both matings and births occur during the wet season in Madagascar, which begins in November and ends in April. That is also the hottest time of the year, so my first thought was that either rainfall or temperature might influence the timing of breeding.

Now, what does this have to do with the question of human reproductive cycles? There is a widespread notion that the normal state for women is regular menstrual cycling, occasionally interrupted by pregnancy. But if you look at primates in the wild, any female is unlikely to have many cycles before becoming pregnant. In fact, the usual condition is pregnancy followed by suckling, with a few ovarian cycles between the end of suckling and the start of the next pregnancy. My studies of lesser mouse lemurs in Madagascar revealed that females generally conceive during the very first cycle of the mating season. After a two-month pregnancy and the birth of her infants, a female may have one more shot at conceiving before the breeding season ends in March. Thus the typical outcome is that each female has, at most, two ovarian cycles and two pregnancies every year. Among higher primates, in the Barbary macaques that I observed on Gibraltar, females of breeding age typically have one or two ovarian cycles and a single pregnancy every year. Like mouse lemurs, most fertile females become pregnant during the first or second cycle of the mating season in autumn. After a six-month pregnancy, they then give birth in spring and do not mate again until the next autumn.

Humans breed throughout the year, so they differ from strictly seasonal primates such as mouse lemurs and Barbary macaques. Nevertheless, in gathering-and-hunting societies without access to contraception women are usually either pregnant or nursing an infant for much of their lives. In these societies women experience only a few menstrual cycles before the next pregnancy begins. Anthropologist Beverly Strassmann's studies of Dogon villagers on the central plateau of Mali, in West Africa, reveal that in a typical lifetime a Dogon woman menstruates about a hundred times. The evidence we have of historical preindustrial societies shows that women in such societies had a life trajectory like that of the Dogon villagers, of successive pregnancies occasionally punctuated by menstrual cycles. In stark contrast, an average woman in a modern industrialized society has notably fewer pregnancies and menstruates about four hundred times in her lifetime. Human females therefore must have evolved to be either pregnant or breast-feeding most of the time. What does this mean for women today?

Contraceptive pills are designed to block ovulation but otherwise simulate the usual pattern of the menstrual cycle. This design is based on the notion that it is natural for women to have long sequences of menstrual cycles when in fact it is far more natural to have many pregnancies with only a few cycles between them. Unlike most other mammals, however, humans can conceive at any time of the year, so women do have a greater potential for having several cycles in a row. With this in mind, contraceptive pills that provoke menstruation from time to time but not in every cycle might be the most effective, in terms of mimicking natural biological patterns.

HUMANS COPULATE AND CONCEIVE throughout the year, but are there any underlying seasonal variations that are not obvious at first sight? Literary evidence, such as Alfred Lord Tennyson's poem "Locksley Hall," tells us that in spring a young man's fancy lightly turns to love. Unfortunately, poetry and folk wisdom fail to tell us whether a young woman's fancy is similarly seasonal. In the scientific literature, substantial evidence indicates that human births usually show a seasonal pattern, with a peak at one time of the year and a clear trough about six months later. Analyses of birth records in the Northern Hemisphere, collected before reliable contraception became commonplace, repeatedly indicated a peak in spring, although exceptions were often noted. In industrialized societies, modern lifestyles affect underlying seasonal patterns. Remarkably, though, evidence for some seasonality of births persists even in such societies.

An annual pattern of births may not tell us much about seasonality of human copulation. It is entirely possible that the probability of conception varies seasonally while the frequency of copulation remains uniform. Be that as it may, Tennyson was seemingly wide of the mark. With nine-month pregnancies, a birth peak in spring corresponds to a peak in conceptions, and perhaps copulations, the previous midsummer.

Adolphe Quetelet, a Belgian polymath who made significant contributions to astronomy, mathematics, statistics, and sociology, was among the first to recognize a seasonal pattern in human births. He is now best remembered for defining the body mass index, or Quetelet index, which is still used with minor modifications in modern studies of human development. In an 1869 treatise on birth and mortality in Brussels, Quetelet presented data from the Netherlands for births during the twelve years 1815 to 1826. His graph showed a clear birth peak in February/March, a deep trough in July, and a minor peak in September/October. Quetelet noted that the seasonal pattern was more pronounced in villages than in towns, which he attributed to differences in environmental temperature. In fact, birth data from the Southern Hemisphere had convinced Quetelet that changes in the position of the sun in the sky were responsible for seasonality. Since Quetelet's time, particularly after the early 1990s, there have been countless reports of variation in human birth rates across the year.

Ellsworth Huntington's 1938 classic compendium *The Season of Birth* was a milestone in investigations of seasonal patterns in human births. Huntington, a geographer with a forty-year connection to Yale University, studied the way humans respond to climatic factors and identified many year-round patterns. He was a leading voice of the environmentalist school, which proposed that human behavior is directly influenced by factors in the physical environment. Huntington was especially interested in how human capacities and achievements varied according to birth month.

A major turning point for biological study undoubtedly came with three papers penned by biologist Ursula Cowgill, all published in 1966. Like Huntington, Cowgill had a long-standing association with Yale. Cowgill examined birth records for human populations around the world and presented conclusive evidence that seasonal patterns are virtually universal. Interestingly, she identified different patterns of peaks and troughs in different regions. Her observations showed that seasonal patterns in the Southern Hemisphere are generally staggered by six months relative to those in the north. Cowgill concluded that annual variation in birth rates is controlled primarily by local climatic conditions, overlain by cultural influences.

Cowgill theorized that ambient temperature might influence conception rate. She also reported that the seasonal pattern had been disrupted to various degrees in urbanized and industrialized regions. One of her studies, based mainly on parish baptism records for the city of York in England between 1538 and 1812, served to illustrate this point. For the first two centuries, from 1538 until around 1752, the pattern matched Quetelet's graph for nineteenth-century Dutch data. Two clear annual peaks emerged, one from February to April and the other from September to November.

From 1752 onward, however, annual variation in births became far less marked, approaching the modern pattern.

Cowgill's seminal papers were followed by abundant publications on human birth seasonality. Particularly notable among these are studies revealing strikingly similar changes in seasonal pattern of births in several European countries over the past century. In an impressively comprehensive paper, published in 2007, a team led by biologist Ramón Cancho-Candela analyzed data for more than 33 million births in Spain over the sixty-year period 1941–2000. They clearly documented a decline and eventual loss of birth seasonality. Initially, a two-peak pattern was evident, with a pronounced birth peak in April and a smaller one in September. After 1970, the pattern began to change, with the peaks becoming less obvious and eventually vanishing altogether in the 1990s. In sum, several studies have shown that European populations originally had a major birth peak in spring and a minor peak in autumn. But this pattern has been weakened and sometimes transformed over the last century.

In 1994, population experts David Lam and Jeffrey Miron concluded that although patterns can change over time, "pronounced and persistent seasonal patterns in births are observed in virtually all human populations." Biologist Franklin Bronson echoed this conclusion and proceeded to examine a range of environmental factors that might be responsible, such as annual variation in temperature and nutrition. He noted that marked seasonal variation in food availability probably influences ovulation. Reduction in food intake or increased expenditure of energy to gather food may delay sexual maturity and reduce ovulation frequency. Seasonally high temperatures in hot climates may also suppress sperm production enough to affect the chances of conception, although this might apply only to men wearing clothes that prevent cooling of the scrotum. High temperatures in the hot season may also suppress ovulation and early embryo survival. On a different tack, it has been suggested that a lower conception rate in the hottest months might be linked to a decrease in sexual activity. In short, seasonal variation in human births has been explained in many different ways.

UNDERSTANDABLY, MANY INVESTIGATORS HAVE TRIED to link seasonal patterns in human reproduction directly to annual variation in specific environmental factors. In seeking an explanation for such patterns, particular care is needed when evaluating indirect evidence from studies of human populations. In all cases, scientists look for patterns in their data, but this is just the first step on a long trek toward real understanding of the processes at work. Ideally, we need experiments to test any ideas about processes, but that is seldom possible when exploring human biology. Lacking experiments, we must be particularly cautious, for claims are often based on very preliminary and inconclusive statistical evidence. As a colleague once aptly remarked: "There are three kinds of anthropologists—those who can understand statistics and those who cannot."

The core problem is that two things may appear to be associated when there is in fact no connection between them. Take, for example, a graph showing the association between brain size and body size across mammals. The pattern we see is that brain size increases along with body size. A statistician would say that brain size is correlated with body size. However, there are several different possibilities when it comes to identifying the underlying cause. The most obvious possibility might seem to be that having a larger body leads to having a large brain. But some scientists have suggested that the brain serves as a pacemaker in development. So perhaps it is the size of the brain that influences body size. There may also be some kind of feedback between brain size and body size, so they increase in concert. Worse yet, it is possible that the observed association between brain size and body size depends on a third factor, some common cause that is not represented in the graph at all. Statisticians call such a common cause a confounding factor.

Pioneering British statistician George Udny Yule, author of the seminal 1911 textbook *Introduction to the Theory of Statistics*, is the reputed source of a famous illustration of a confounding factor at work. Pleasingly, his illustration comes from the realm of reproduction. Yule noted that in villages in Alsace in eastern France the number of human babies born is correlated with the number of storks nesting locally—the greater the number of storks, the greater the number of babies delivered each year. Although it might be tempting to see this as evidence that storks do actually deliver babies, the real explanation is much more mundane. Larger villages have more houses with chimneys on which storks can build nests, and on average more babies are of course born every year in larger villages. The confounding factor here is village size. In this case, as elsewhere, a comparative approach is helpful. Storks have a limited geographical distribution, so all we need to do is to examine human birth rates in villages that naturally have no storks. If any babies are born at all in storkless villages, the notion of a causal connection between storks and babies collapses.

The field of human reproduction is littered with examples of interpreting mere association as evidence for causation. A simple misconception (no pun intended) arises from the observation that women who do not menstruate do not become pregnant. This led to the early belief, which persisted until the 1930s in Western society, that menstruation and conception have a direct causal connection. It was originally thought that conception results from mingling semen with menstrual blood. Accordingly, menstruation was believed to be the fertile time in a woman's cycle. This view persisted long after it was realized that a sperm must fertilize an egg, because ovulation was wrongly thought to coincide with menstruation as well. As a result, for some considerable time women were instructed to avoid copulation during menstruation and to treat midcycle as the "safe period"—the exact opposite of advice given after the 1930s. To take another, more subtle example that is directly connected with seasonality, one study showed that human conceptions and sales of contraceptives both peaked in the summer. Whereas a direct interpretation might be that contraceptives actually increase the likelihood of conception, this is merely an example of the general rule that things that vary over time may show coincidental patterns.

NUMEROUS INVESTIGATORS, INCLUDING QUETELET, Huntington, and Cowgill, have tried to link seasonal patterns in human breeding to annual variation in environmental factors. Temperature is a popular candidate, because men's testes are so sensitive to heat. However, in any given region the annual pattern of variation in temperature remains fairly stable from year to year, so environmental temperature cannot explain changing patterns of birth seasonality. A similar argument applies to rainfall, another much-cited factor. Yet there is an alternative possibility. Perhaps seasonal patterns of breeding developed over evolutionary time to match average annual variation in environmental conditions. In that case, the breeding pattern may be driven by internal factors rather than responding to environmental conditions encountered at any particular time. As physiologist Alain Reinberg has noted, rhythmic activity is a basic property of living matter, from single-celled organisms to humans.

As it happens, I had a golden opportunity to learn about internal biological clocks at a time when research was really getting under way on the topic. When I was a graduate student in the mid-1960s, studying the behavior of treeshrews at the Max Planck Institute in Seewiesen, Germany, down the road, nestled in the village of Andechs, was a sister institute dedicated to pioneering research on the internal clocks that animals use to keep track of time. The most basic type of internal timekeeper, present even in single-celled organisms, controls activity patterns and biological processes over each day/night cycle. This circadian clock marks off a period of about twenty-four hours. The approximate timing provided by the internal mechanism is fine-tuned by cues from the environment, most notably the presence or absence of daylight.

The Andechs research institute was led by physician and physiologist Jürgen Aschoff, who was one of the founding fathers of research into internal clocks, now known as chronobiology. He and a team of researchers, including ornithologist Eberhard Gwinner, were conducting a series of experiments on mammals and birds to explore the workings of the day/night clock. One of the most striking findings of this early research was that when cues from the environment are excluded, animals will show internally generated activity cycles that can vary by up to a few hours on either side of the standard twenty-four-hour period. For instance, an animal might have a free-running cycle with an average duration of about twenty-six hours. In other words, cues from the environment are needed to constrain the clock to run with a regular twenty-four-hour rhythm. It is rather like a grandfather clock that does not keep very good time and must be reset every dawn and dusk to nudge it back on track. Anyone who has suffered from severe jet lag knows that we pay a heavy price for messing with our internal clocks by switching time zones and thus radically shifting those all-important light cues from the environment.

Research at Andechs was not limited to animals. After experimenting on himself, Aschoff recruited individual student volunteers to live for up to four weeks at a stretch in a specially constructed underground bunker from which all external time cues were rigorously excluded. Every volunteer surrendered all timepieces when taking up residence in the bunker and was left to regulate activity independently by switching lights on and off at will. Volunteers prepared their own food so that there were no external time clues from catering staff. Each volunteer also received a daily allowance of one bottle of the extra-strong Andechs beer brewed by a local monastery. Under isolation in the bunker, everyone's internal circadian clock ran free. As with animal studies, the length of the sleep/wake cycle was usually found to differ from the twenty-four-hour standard by up to a few hours. The average duration of a free-running human internal clock in Aschoff's bunker was about twenty-five hours.

In addition to the twenty-four-hour clock, long-lived animals and plants have another internal clock that marks off the passage of a year. Time cues from the environment are also essential to keep this circannual clock on track. In many cases, annual variation in day length has been identified as the key factor for fine-tuning the internal circannual clock. Day length, the period between sunrise and sunset, varies in a systematic, predictable fashion over the year and can serve as a reliable cue to time across seasons. But using day length as a cue for a circannual clock has one potential limitation: The amount of variation depends on latitude. At high latitudes, day length varies by several hours across the year, whereas at low latitudes, close to the equator, annual variation is barely noticeable. In the Northern Hemisphere, the longest days are in late June and the shortest in late December. In Chicago, for instance, the longest day in midsummer lasts more than fifteen hours, while the shortest day in midwinter is close to nine hours, giving a maximum difference of over six hours across the year. On the equator, by contrast, day length varies by only a few minutes over a year, so in tropical regions it is likely to be far less effective as a time cue for annual cycles. Although less experimental work has been conducted to explore circannual clocks compared to circadian clocks, because studies must last years rather than months, it has been repeatedly shown with animal experiments that removing the environmental cue of changing day length commonly results in free-running, internally generated cycles that are approximately a year in length.

Seasonal breeding is extremely widespread among mammals, ranging from strictly limited birth seasons to year-round breeding with a moderate birth peak at some stage. In many species, times of mating, conception, and birth are linked to the regular annual pattern of variation in day length. In such instances, both the development of testes in males and sexual activity in females are triggered by a particular phase of the annual day length cycle. The most obvious effects of control by day length are, of course, found in species with tightly bounded birth seasons. Among primates, this is most evident in lemur species living on Madagascar. In a few instances, experimental manipulation in the laboratory has yielded direct evidence of an influence of day length on breeding. In my own work at University College London, for example—benefiting from that earlier research on biological clocks at Andechs—I programmed lesser mouse lemurs to breed at a suitable time of the year with a special light-clock that automatically generated a varying day length pattern resembling that in Madagascar. Increasing day length triggers the mating season in mouse lemurs, so I simply set the light-clock to stimulate breeding at a time of the year when it was convenient for me to conduct observations. At one point I even managed to reduce the interval between breeding seasons by compressing the annual day length cycle into nine months.

As it happens, experiments lasting several years are not needed to find out whether reproduction in a mammal species is likely to be triggered by annual variation in day length. There is a shortcut arising from the natural consequences of Earth's rotation around its inclined axis while orbiting the sun. The annual pattern of day length variation in the Southern Hemisphere is a mirror image of that in the north, such that, in the south, the shortest days are in late June and the longest in late December. Because of this, transferring day-length-dependent mammals from one hemisphere to the other will shift their breeding seasons by six months. The breeding records of zoos, which often house mammals transferred between hemispheres, provide useful information here.

However, variation in day length is simply an external cue for the time of year; it does not determine when an animal is most likely to breed. Mating, pregnancy, birth, and suckling have all been proposed as the primary factor driving the timing of the breeding season. But my own study of seasonal breeding in lemurs revealed that annual patterns of mating, birth, and pregnancy differ markedly between species. Larger-bodied lemurs have longer pregnancies and nurse for longer than small-bodied lemurs. The biggest lemurs mate, complete pregnancy, and even begin suckling during the dry season, when food resources are limited and day lengths are short. The only common factor that I could identify was this: Every species gives birth at the appropriate time for infants to be weaned and then feed independently to accumulate extra resources before the wet season ends. This permits offspring to survive the hardships of the following dry season. Achieving independence before a seasonal period of food shortage may well be the key to offspring survival. Examination of other primates shows that this explanation generally fits all of the patterns observed. Of course, with large-bodied species such as apes and humans that are adapted to nurse for years rather than months, the link between birth seasonality and food resources is far less clear.

THERE IS MUCH TO BE LEARNED about seasonal reproduction by examining our primate relatives. For many decades, the rhesus monkey, a red-faced denizen of Asian forests, was the standard laboratory primate for medical comparisons with humans. Rhesus monkeys have a huge geographical range in Asia, from eastern Afghanistan and northern India across to southern China and Thailand. At one time, they were easily available for research and were imported into the United States and Europe in great numbers. Studies of reproduction in this species once provided the bulk of findings obtained from nonhuman primates to interpret human reproductive biology. For many years, "nonhuman primate" essentially meant "rhesus monkey."

More than anyone else, biologist Carl Hartman pioneered the use of rhesus monkeys as a model for human reproduction. His monograph on their reproduction, published in 1932, is an enduring classic. In 1938, soon after Hartman had begun his work with a laboratory colony, some four hundred rhesus monkeys were released on the uninhabited thirty-eight-acre island of Cayo Santiago, off the southeastern coast of Puerto Rico. This provisioned but free-ranging colony has been continuously maintained ever since, yielding a wealth of information on behavior and reproduction of free-ranging rhesus monkeys.

The founding stock for the Cayo Santiago colony was collected in northern India by psychologist Clarence Ray Carpenter, a pioneer in field studies of primates. The rhesus monkeys on Cayo Santiago rapidly organized themselves into groups whose descendants have been closely monitored ever since. In 1942, Carpenter reported that the Cayo Santiago monkeys bred seasonally, although it took some years for them to settle into a regular pattern. During the mating season, adult females showed a series of menstrual cycles, each including a receptive period lasting about nine days. Citing animal dealers in Calcutta, Carpenter noted that rhesus monkeys also had a three-month birth season in their natural habitat in India, a fact since confirmed by long-term field studies.

The free-ranging rhesus monkeys on Cayo Santiago also show a seasonal cycle in reproductive characteristics of both males and females. In 1964, anthropologist Donald Sade discovered that the testes are largest during the mating season and considerably smaller during the birth season. The next year, together with Clinton Conaway, he showed that sperm production also has an annual cycle, peaking during the mating season in the fall. Other studies showed that in rhesus monkeys female cycles are limited to the period from July to January, with intensive mating occurring in September and October, and this pattern continues today on Cayo Santiago. Overall, then, rhesus monkeys kept in free-ranging colonies or outside enclosures show seasonal breeding quite similar to that in their Indian homeland.

It is important to know whether a seasonal pattern persists in a controlled laboratory environment. Indeed it does. Carl Hartman reported in 1931 that marked seasonal variation in the conception rate persists and that ovulation almost completely stops for part of the year. Reproductive biologists Richard Michael and Barry Keverne confirmed that a similar birth peak in March and April had been seen in various laboratory colonies since Hartman's time. In their own colony, they observed a high ejaculation frequency from November to January, with a prominent peak in December, contrasting with low levels between February and May. This persistence of a seasonal pattern in captivity clashes with the interpretation that environmental factors directly drive mating. Still, it is important to note that the seasonal pattern reported by Michael and Keverne could have been influenced by variation in exposure to daylight. Although ambient temperature was maintained approximately constant during the study, a standard fourteen-hour period of artificial lighting was augmented in summer by natural daylight, which increased day length by up to two hours between mid-June and mid-July.

Exposure of the rhesus monkeys studied by Michael and Keverne to variation in day length across the year may have influenced the breeding pattern. In fact, in 1932, the astute Hartman had already noticed that zoo curators in Australia reported a six-month shift in the birth season compared to the Northern Hemisphere. To test this further, psychologist Craig Bielert joined forces with John Vandenbergh to study annual birth patterns in zoos in New Zealand and South Africa as well as Australia. They predicted that a six-month difference in the birth peak should accompany the shift in hemisphere. And they were proven right, showing that mating in the south was largely confined to the period from March to August and that most births occurred between October and January.

But what happens when day length remains completely unchanged across the year? Reproductive biologist Jean Wickings and her colleague Eberhard Nieschlag addressed this question in a study of adult male rhesus monkeys that had been isolated from females for up to four years in a controlled laboratory environment with no seasonal changes in light, humidity, or temperature. Testis size, internal testis structure, sperm production, testosterone levels, and ejaculation frequency all showed a marked annual pattern, peaking in the autumn and winter months. This persistence of a distinct seasonal pattern in rhesus monkeys isolated from any external environmental influences confirms that an internal regulatory mechanism exists. Thus seasonal breeding in rhesus monkeys is governed by an internal annual clock that is fine-tuned by changes in day length.

IS THERE ANY PARALLEL TO HUMANS, given that wild-living rhesus macaques have a well-defined breeding season rather than just a seasonal peak in year-round breeding? As Cowgill reported, birth data from human populations in the Southern Hemisphere show a pattern six months out of phase with that reported for the north. Whereas the birth peak commonly occurs in spring in the Northern Hemisphere, the peak in the south generally coincides with autumn in the north. This suggests that human breeding seasonality is linked to variation in day length, as in rhesus macaques. This clue is reinforced by the fact that seasonality of conceptions and births is reduced or absent in human populations living near the equator, where there are only minor changes in day length over the year. One worldwide study showed that human births become increasingly seasonal at higher latitudes. The correlation of seasonality with increased latitude seems to rule out the possibility that temperature is the main driver. While the hottest time of the year in the north is six months out of phase with the hottest time of the year in the south, the degree of birth seasonality is most marked at high latitudes, not near the equator, where ambient temperatures are typically the hottest.

Of course, it is out of the question to test the mechanism underlying human seasonality by keeping human subjects under controlled laboratory conditions for several years. So we will always have to resort to circumstantial evidence to seek an explanation for seasonal patterns in human reproduction. And these patterns do exist: Various studies have revealed variation in sperm quality and testosterone levels in men over the year, and seasonal differences have been detected in the time of day at which women ovulate.

In an attempt to tease apart the biological and social factors that might influence annual patterns in human births, Jürgen Aschoff conducted a sophisticated statistical analysis together with Till Roenneberg, a medical psychologist. In an analysis of a data set for 166 regions around the world, with a total of 3,000 years of monthly birth rates, Roenneberg and Aschoff proved that the observed pattern depends on latitude, with a six-month difference between the Northern and Southern Hemispheres, and that the degree of seasonality increases with latitude. The results, published in two papers in 1990, demonstrated for the first time on a global scale that variation in day length does influence reproduction in humans, as it does in many other animals. Roenneberg and Aschoff concluded that although social influences may influence the timing of conceptions, seasonality of human reproduction is driven primarily by biological factors. Ambient temperature is also a significant factor. At temperatures between 40°F and 70°F, conception rates exceed the annual average, whereas decreased conception levels are found with extreme temperatures above or below that range.

Because conception time seems to be a key aspect of human birth seasonality, it raises the question of whether copulation and characteristics of sperms or eggs also show seasonal variation. Is there an optimal time of the year for sex cells? There is, but—surprisingly—sperm concentration seems to be minimal in summer, when sexual activity is at its peak. In 1984, for instance, reproductive epidemiologist Alfred Spira reported on a study of more than 1,000 ejaculates collected from fifty-two New York medical students for up to three years. Two peaks were found in semen volume, concentration, and total sperm count, one at the end of winter and beginning of spring and the other in late autumn. By contrast, a peak in the percentage of normally formed sperms and sperm motility was found in late summer, whereas a trough occurred in late winter and early spring. These results indicate that although sperms are less abundant in late summer, their condition may be optimal then, and it may be poorest in late winter and early spring, when sperm numbers peak. Spira reviewed several other studies that generally pointed in the same direction. Production of optimal sperms in late summer would fit well with a birth peak in spring, but it is surprising that sperm numbers are low rather than high in late summer.

Studies of assisted reproduction are particularly illuminating for examining seasonality, as they provide the closest thing to a controlled laboratory environment as is possible for human reproduction. In a 1988 study, gynecologist Eftis Paraskevaides and colleagues observed a seasonal pattern in more than 250 conceptions resulting from artificial insemination by donors in their clinic. Conception was found to be more common from early winter until early spring (October to March), with a peak in November. As sperm counts had been found to peak at the end of that period, in February/March, this result suggested some seasonal variation in the quality of the egg or in the receptivity of the inner lining of the uterus.

Seasonal variation in the success of in vitro fertilization (IVF) was examined by gynecologist Simon Wood, working at Liverpool Women's Hospital. In a 2006 paper, he and his colleagues analyzed almost 3,000 standardized cycles of IVF combined with intracytoplasmic sperm injection and found a significant improvement in the outcome when performed in the months with extended day length (April–September) compared to months with shorter day length (October–March). During months with longer day length, ovarian stimulation was more effective, implantation rate per embryo transferred was significantly improved, and the clinical pregnancy rate was increased from 15 percent to 20 percent. However, the rate of fertilization in vitro showed no significant difference between summer and winter. Differences were even more pronounced when long-day-length months (April–September) were compared with short-day-length months (October–March) for patients treated at both times of the year.

FOR PEOPLE LIVING IN industrialized societies, seasonal patterns in human births might seem to be no more than an intriguing heritage from our biological past, with little relevance today. But recent reports suggest that seasonal births may be linked to human health. To take one striking example, in 1953 pediatrician Gregor Katz reported a seasonal pattern in the incidence of premature births. Full-term births in the Swedish population, following the typical European pattern, showed a peak in spring. By contrast, Katz found a peak in January, about two months earlier, in a sample of more than two hundred premature babies born between 1944 and 1951 at a Karlstad hospital. The premature babies were born, on average, about one and a half months too early, so the peak for their conception would more or less coincide with the conception peak for full-term babies. In other words, a summer peak in conceptions fits the birth peaks for both premature and full-term babies. A more recent study by human ecologists Shinya Matsuda and Hiroaki Kahyo used a sophisticated statistical analysis to examine seasonal variation in premature births. They looked at data for more than 7.5 million births in Japan in the five-year period from 1979 to 1983 and found a distinct peak of conceptions leading to premature birth occurring in May and June, corresponding to late spring and early summer. This would result in a peak in premature births in December and January, thus matching the pattern identified by Katz with a far smaller sample.

As Katz astutely noticed, it is important to bear in mind seasonal fluctuation in the percentage of premature babies when considering other findings. Such fluctuation may partly explain the relationships between birth month and other features. The shorter pregnancies leading to premature births are reflected in lower birth weights, which might in turn account for the intriguing seasonal differences in intelligence and achievement reported by various authors, such as Ellsworth Huntington.

A different perspective is provided by a 1960 study of more than 1,000 miscarriages, conducted by obstetricians Riley Kovar and Richert Taylor at a hospital in Omaha, Nebraska. They found no clear seasonal pattern, but the miscarriages occurred in clusters. In addition, there was a direct relationship between numbers of miscarriages and daily temperature changes. So once again it seems that both annual changes and ambient temperature interact to influence human reproduction. There are doubtless other factors that come into play, among which seasonal variation in infection is a strong candidate.

A LARGE BODY OF EVIDENCE now indicates that annual variation in day length influences reproduction in human populations, especially those at medium to high altitudes. However, observed seasonal patterns in humans have generally weakened over the past century, and in some cases the annual birth peak has shifted. It seems that the influence of ambient temperature has increased over that period. It is likely that altered living conditions—corresponding with increasing industrialization and the greater use of artificial lighting—have reduced the effect exerted by natural variation in day length.

Monitoring changes in day length is one of the rare areas where it has been possible to conduct experiments with human subjects. A team led by clinical psychobiologist Thomas Wehr made several striking findings regarding exposure to light and biological rhythms in human subjects. In one 1991 study, Wehr tested the effects of artificial changes in day length on sleep duration and secretion of the hormone melatonin. Melatonin, known as the "hormone of darkness," is secreted into the blood of mammals by the tiny pineal gland in the brain, a remnant of a third eye that was once present in the skull roof of early reptiles. The human pineal gland is about the size of a grain of rice. Melatonin is secreted only during the hours of darkness and plays a direct part in coordinating biological clocks. This is why melatonin is sold as an over-the-counter remedy to counter effects of jet lag.

In Wehr's study, eight volunteers were exposed to a "summer" schedule of sixteen hours of light and eight hours of darkness for one week and to a "winter" schedule of ten hours of light and fourteen hours of darkness for four more weeks. As in other mammals, the duration of nighttime melatonin secretion in the volunteers was significantly longer (by more than two hours) after exposure to the winter schedule with short day lengths. The duration of the sleep phase was also longer. Wehr had previously found that the setting of a subject's biological clock persisted during a twenty-four-hour test period in which a volunteer remained continuously awake in constant dim light. Differences between summer and winter schedules were confirmed under these test conditions, revealing that prior exposure to a particular day length resets the internal clock in a lasting manner.

Wehr also investigated the effects of seasonal changes in day length on associated hormonal patterns. In response to light, the twenty-four-hour biological clock synchronizes the timing of the biological day and night so that they appropriately match the timing and duration of the solar day and night. At the same time, it adjusts the duration of the biological day and night to track seasonal change in lengths of the solar day and night. Wehr found that, in addition to influencing nighttime melatonin secretion, changing day length affects the production of other hormones by the pituitary gland, such as prolactin, cortisol, and growth hormone.

Quite apart from day length or the duration of daylight hours between dawn and dusk each day, light itself may influence human biology. Wehr has noted parallels between seasonality in human reproduction and seasonal affective disorder (SAD), suggesting that common biological processes might be at work. SAD is a medical condition, first formally recognized in 1984, in which episodes of depression occur at a particular time of the year. Symptoms most often occur during winter, but SAD may arise at other times of the year. This variation explains why the disorder has multiple names, such as winter depression, winter blues, summer depression, or summer blues. Symptoms include increased sleep and daytime sleepiness, and predisposing factors include the amount of ambient light and levels of certain hormones. The risk of SAD is greater for people living in places with long winter nights, and the condition occurs more often in women than in men. Cloud cover may also contribute to negative effects of SAD. People generally experience a shift to lower energy levels in winter, when days are shorter, but SAD is a more severe condition regarded as a category of depression.

Treatments for the winter form of SAD include light therapy, using natural sunlight or bright fluorescent lights, and carefully timed administration of melatonin. In therapy with an artificial bright light, the patient sits one to two feet away from the light source for up to an hour at a time, without looking directly at it. This is usually done in the early morning, to imitate sunrise, and simulation of dawn has proven to be particularly effective, again suggesting a connection with the biological clock. Another type of light therapy involves increased exposure to sunlight, with the SAD sufferer either spending more time outdoors or using a computer-controlled heliostat to reflect sunlight into the windows of a living space. However, light therapy does not work for everyone. Only a quarter to half of patients diagnosed with SAD gain appreciable relief. Some scientists have attributed SAD to misalignment of sleep/wake alternation with the biological clock, and treat it by administering melatonin in the afternoon.

Ovaries and testes both have binding sites for melatonin, suggesting a direct link between biological clocks and the production of sex cells. In 2007, chronobiologists Konstantin Danilenko and Elena Samoilova showed that exposure to light affects reproductive hormones and ovulation in women, using trials to assess the stimulatory effect of morning bright light. Their work followed studies revealing that menstrual cycles become shorter when women with abnormally long cycles or suffering from winter depression are exposed to light. Twenty-two women were followed over two separate menstrual cycles, either in their own homes or at medical centers. In one of the cycles, bright light was administered during the follicular phase for forty-five minutes shortly after awakening every day for a week. In the other cycle, the same procedure was followed but with dim light. Prolactin, luteinizing hormone, and follicle-stimulating hormone were all significantly increased during cycles with exposure to bright light. Follicles growing in the ovary were also larger and the ovulation rate was higher. Such a direct link between exposure to light and the progress of the ovarian cycle may explain why women are especially susceptible to SAD.

Regardless of the underlying mechanism, it is clear that natural daylight plays a major part in human reproduction. Historical and experimental evidence indicates that human responses to seasonal changes in day length were more pronounced before the industrial revolution. One outcome of industrialization has been that these responses have been increasingly suppressed by changes in our living conditions. This leads on to a question: Do continuously available artificial light and repeated changes in time zones through international travel adversely affect our natural reproductive mechanisms? SAD provides one example of the ways in which our health may be affected. Regardless, it is unlikely that industrialized societies are likely to do much about it, much less limit the use of artificial light. A commentary by Natalie Angier published in the *New York Times* in 1995 quoted Wehr as saying: "We're addicted to our endless summer."

While such language may be on the fanciful side, serious issues are at stake. Natural production of melatonin declines across the life span, hence the increases in sleep disruption with age. Surveys indicate that melatonin has beneficial effects for sleep in old age and perhaps for various age-related diseases as well. More alarming, melatonin levels have also been linked to cancer. Women with cancer of the womb lining have been reported to have significantly decreased melatonin levels. In patients with breast cancer, melatonin levels are lowered even more drastically, by as much as 90 percent. As is well known, the incidence of breast cancer has increased markedly, almost tripling since the 1940s in the United Kingdom, for instance. Hitherto unidentified environmental factors are doubtless responsible. Incidental findings, such as a reported association between night shift work and an increased risk of breast cancer, suggest another possible link between cancer and lack of melatonin. Yet we must be cautious, as causation does not automatically follow from correlation. Melatonin may decline as a consequence of cancer and age-related diseases in general, not the other way around. Nevertheless, we would do well to consider the possibility that radical disruption of natural light conditions in modern lifestyles may have an adverse impact on melatonin levels, affecting our reproductive lives and, worse, opening the way to degenerative diseases.

CHAPTER 3

From Mating to Conception

Spindly-armed, tailless gibbons are a common sight in zoos, swinging elegantly from branch to branch with their zoobound brethren. In their natural habitat in the forests of Southeast Asia, these small apes, most weighing around twelve pounds, typically live in small family groups. Each family, containing a single adult pair and any growing offspring, occupies a well-defined territory. In most species, an adult pair advertises its territory with melodious "duets" accompanied by refrains from other group members. Adult pairs generally remain stable for many years, but offspring leave their family groups once they are sexually mature to seek partners and establish their own territories.

Gangling orangutans, the largest tree-living mammals, inhabit some of the same forests as the gibbons. Doubtless because of their size, orangutans are slow-moving, cautious climbers and lack the spectacular acrobatic skill of gibbons. Orangutans of both sexes are big, but there is a striking disparity between adult females, which weigh around 90 pounds, and adult males, averaging about 150. Such marked sexual dimorphism, as it is known, contrasts starkly with the roughly matched body weights of male and female gibbons. Orangutans also differ from gibbons in their social organization, as they are a largely solitary species. (I always delight in telling students that the average group size is one and a half orangutans.) Adult males are almost exclusively solitary. As a male matures, he develops distinctive cheek flanges and begins to utter loud, booming calls that make known his presence in an extensive territory. This area commonly encompasses several smaller territories occupied by individual adult females. Social tendencies are shown only by females, which may associate with their offspring for some time, and by subadult individuals, which sometimes literally hang out together. Encounters between an adult male and an adult female within his range are generally confined to brief mating episodes. Any local subadult male will usually avoid the adult male but may rapidly mate with an adult female. Such sneaky matings are often forcible, leading some fieldworkers to describe them as rapes.

Africa has no small-bodied apes, but it is home to two different kinds of great apes: gorillas and chimpanzees. Both are at least partly ground-living. Gorillas, like orangutans, show pronounced sexual dimorphism. An average adult female weighs about 200 pounds, whereas an adult male is considerably heavier, around 330 pounds. Males are also patterned differently, as hairs on their lower back turn silvery gray at adulthood, earning them the label "silverback." The social organization of gorillas, meanwhile, is unlike either the orangutan or gibbon structure. As a rule, an established gorilla group contains a single silverback male along with several adult females. Essentially, then, gorilla groups are "harems," although immature males, known as "blackbacks," and occasionally additional silverbacks may be present. Mating within a group is largely restricted to the resident silverback.

Chimpanzees, the other African great apes, contrast with gorillas in several ways. For starters, body size differs only mildly between the sexes, with adult females weighing about seventy-five pounds and adult males around ninety. Aside from this limited difference in size, males and females are not very different. More important, the social system of chimpanzees differs from that of all other apes. Indeed, it is so complex that it took years of field study to reveal its various levels of organization. In the wild, common chimpanzees are usually seen in small feeding groups of varying composition containing perhaps half a dozen individuals. So the social life of chimpanzees at first seemed to be relaxed and flexible. However, it was eventually discovered that members of temporary feeding groups actually belonged to a large social unit containing about eighty individuals, including adults of both sexes. Flexibility exists only within the unit. A social unit of chimpanzees is rarely seen moving around as a cohesive cluster, but it is an important higher level of organization. Units are territorial and clashes between them can be lethal. Within a unit, mating is generally promiscuous, with each receptive female mounted by several males. But here, too, complications exist. Sometimes an individual male and female will wander off on a private "safari," avoiding promiscuous mating. Furthermore, common chimpanzees differ in important respects from the closely related bonobos, so-called pygmy chimpanzees, which are actually just rather slender and only marginally smaller.

This brief outline of social patterns in apes, our closest zoological relatives, reveals two important points. First, a basic mating pattern emerges for each kind of ape—monogamy in gibbons, harems in gorillas and orangutans (although the latter are dispersed rather than group-living like the former), and promiscuous multimale-multifemale groups in chimpanzees. The second point is that social systems can differ widely even within a group of relatively closely related primates such as the apes.

Although patterns of social organization vary widely across primates, we can generalize three basic categories, as illustrated by the four species above: monogamy, polygyny (or harems), and promiscuity. In principle, there could also be a fourth kind of group in primates—polyandrous groups containing a single adult female with several adult males and youngsters. This fourth kind of organization is exceedingly rare among primates.

Field studies have revealed that each group-living primate species has a typical pattern of social organization: pair-based, harem-living, or multimale-multifemale (which I shall simplify to "multimale"). Although a few species show some variation, usually alternating between harems and multimale groups, most show a stable pattern. It is reasonable to assume that species-typical social patterns have some genetic basis. In captivity, primates typically show the same pattern as in the wild. For example, keeping gibbons in zoos is successful only if adults are caged as pairs, along with any growing offspring. On the other hand, field observations have revealed that closely related species—which are of course genetically very similar—can show distinctly different social patterns. In Madagascar, for instance, mongoose lemurs live in pair-based groups, while closely related brown lemurs live in multimale groups. Similarly, in Africa plains baboons live in multimale groups, whereas sacred baboons live in harem groups. This reinforces the idea that social organization can evolve quickly.

Setting out from this general background, we can go on to human patterns of social organization and reproductive choices. Cultural influences are so strong that it is by no means obvious what is "natural" for our own species when it comes to choosing mates. While we can turn to general principles derived from comparisons with other primates, we must remember that social organization is flexible and can evolve rapidly.

Humans, like all other primates that are active during the day, are gregarious and live in recognizable groups. Of course, the complexity of modern societies greatly exceeds that of any nonhuman primate grouping, but much of that complexity dates back little more than 10,000 years, to the beginnings of settled communities with domesticated plants and animals. For at least 99 percent of the time since our lineage diverged from the chimpanzee lineage, we lived as roaming gatherer-hunters in relatively small bands.

If we try to classify human social organization using the broad categories identified for nonhuman primates, we at once hit a problem. Modern human societies cannot be limited to one category. Taken as a whole, human societies show virtually all of the basic patterns: some are monogamous, while others are polygamous, including rare cases of polyandry. However, no human society seems to be genuinely promiscuous in the vein of chimpanzees. Nevertheless, human social organization is unusually flexible. Emerging flexibility—reflecting weaker biological constraints—was doubtless one hallmark of our evolution. In their 1951 classic *Patterns of Sexual Behavior*, reproductive biologists Clellan Ford and Frank Beach reviewed information for almost two hundred human societies, concluding that polygyny predominated, occurring in three-quarters of the sample. But we cannot say that polygamy was necessarily ancestral for humans, for even in "polygamous" societies, unions are often monogamous by default because many men lack the resources to have multiple wives. Some authors, such as Desmond Morris in *The Naked Ape*, have concluded that monogamy is the dominant pattern because of sheer weight of numbers in certain societies today. Ultimately there has been little indication of any inherited biological basis for either monogamy or polygamy in humans.

Comparisons between humans and our closest relatives, the apes, also fail to yield a clear answer. Apes show all basic patterns identified among primates generally, although, because chimpanzees seem to be our closest relatives among the apes, it is often assumed that a similar multimale, promiscuous social system was the starting point for human evolution. Yet this conclusion is a prime example of "frozen ancestor" thinking, in which one living species is taken as a model for the origin of another. As discussed above, social patterns evolve rapidly and can differ markedly between closely related primate species. Wide variation among apes underscores that point. Thus we cannot simply assume that the common ancestor of humans and chimpanzees behaved like modern chimpanzees.

Pair-based groups are in a minority among nonhuman primates, found in only 15 percent of species. The remaining 85 percent live in single-male or, less commonly, multimale groups containing several females. When it comes to group-living mammals other than primates, most (apart from a few exceptions, such as lions and certain other carnivores) are polygynous. They live in single-male harem groups, and surplus males often cluster in bachelor groups. Pair formation is even rarer, occurring in only 3 percent of nonprimate mammal species. By contrast, most birds—about 90 percent—are pair-living. One might aptly say: "Higamus, hogamus, birds are monogamous; hogamus, higamus, mammals polygamous."

Why do most birds live in pairs whereas most mammals do not? The likely explanation is that birds are obliged to share parental care. In pair-living birds, the male commonly helps incubate and feed the young so that the female can leave the nest to forage. Mammals lack such constraints. In marsupials and placentals, offspring develop within the mother's body until birth, the equivalent to hatching in birds. Following birth, a mammal mother draws on her own body reserves to suckle offspring instead of foraging for food morsels. Because of this arrangement, parenting by male mammals is not usually obligatory. Sure enough, in most mammal species males make no direct contribution to rearing offspring. A comparison with birds suggests that the rare evolution of monogamy in mammals may be linked to paternal care. Behavioral biologist Devra Kleiman showed that this is the case for several primates and various other mammals, especially members of the dog family. It is notable that human babies are particularly helpless compared to those of other primates and therefore require intensive parental care. As I'll explain in Chapter 5, this helplessness reflects a special development in human evolution. The upshot is that, in comparison with other primates, social support is crucial for development of our infants.

* * *

It is tempting to assume that mating arrangements directly match the typical social pattern. For instance, in a pair-living species it might seem obvious that the adult male sires any offspring of the pair. In other words, pair-living social organization and strictly monogamous mating may appear to be two sides of the same coin. Take, for example, the thousands of pair-living bird species that have traditionally been regarded as strictly monogamous. Countless hours of observation by dedicated bird-watchers failed to reveal any deviation from strict limitation of mating to the partners in a pair. However, the advent of DNA-based paternity tests changed all that. Surprisingly, for nine out of every ten bird species studied, it was discovered that the pair male did not consistently sire all offspring in a nest; about half of them resulted from extrapair copulations. How, we may ask, did such extrapair mating escape detection through the binoculars of all those eager bird-watchers? The answer is that "sneaky" copulations are rapid and discreet. The pair male may be as unaware of them as the peeping ornithologist.

Theorists have leapt in to explain these exciting new findings, incautiously borrowing loaded terms from the human arena. Ironically, the Middle English "cuckold," used to describe human infidelity, was originally derived from an old French word for cuckoo. Coming full circle, it is now widely applied to animals to refer to pair-living males tending offspring that may not be their own. The customary explanation for this occurrence is that it is in a female's genetic interest for different males to father her offspring. She may pair up with one male to ensure paternal care while also engaging in extrapair copulations that will boost genetic variability in her offspring. It is in a female's interest to conceal from her pair mate any sneaky copulations with other males. If the female's partner were to become aware of the threat to paternity, he might abandon the nest. Or so the argument goes.

This explanation assumes that a male's overriding interest is to strive for exclusive paternity of any offspring in his nest. Hence strong selection would be expected for any mechanism that reduces the likelihood of extrapair copulation. Yet chances are that any pair-living male will himself seek mating opportunities with females of nearby pairs. As is the case for females, it is surely in a male's genetic interest to father offspring in nests of females paired with other males. Perhaps there is a trade-off between a male's interest in defending paternity of offspring in his own nest and benefits from contributing to paternity of offspring in other nests. Males, too, can spread their options. In any event, the mating arrangements of pair-living birds have turned out to be considerably more tangled than originally believed.

Thus patterns of social organization and mating systems are not simply two sides of the same coin; they can vary independently, at least to some extent. This variation also applies to mammals, including primates. DNA-based tests have been used to assess paternity in a few mammal species, and similar results have emerged. Compared to the wide array of studies conducted with birds, there have been few genetic studies of paternity in pair-living mammals, but evidence for frequent extrapair copulations will doubtless emerge. To take one example, in 2007 behavioral biologist Jason Munshi-South conducted a DNA-based study of a pair-living treeshrew species in Sabah (Borneo) and found a high level of extrapair paternity. Similar evidence has also been reported for certain pair-living primates—for example, for fork-crowned lemurs and fat-tailed dwarf lemurs, both night-active inhabitants of Madagascar. Even gibbons, the prime example of monogamy in primates, have been observed engaging in sneaky extrapair matings in the wild.

Social patterns and mating systems can differ sharply in primates. Verreaux's sifaka, a day-active Madagascar lemur, typically lives in small, multimale groups containing about half a dozen adults of both sexes, although the male-to-female ratio varies greatly. As in most other lemurs, mating in Verreaux's sifakas is strictly seasonal, confined to a few weeks each year. Pioneering fieldwork by biological anthropologist Alison Richard revealed that the mating season is a time of turmoil. Confirming earlier anecdotal accounts, she found that severe fighting often erupted during the mating season. Moreover, she observed that mating typically occurred between groups rather than within them. Reorganization of social groups often ensued. Thus group structure does not dictate mating arrangements and evidently has other functions. Indeed, primate field studies indicate that social organization is generally linked to feeding. So while Verreaux's sifakas live in stable feeding groups for most of the year, those groups are disrupted by seasonal mating.

THE DISTINCTION BETWEEN SOCIAL ORGANIZATION and mating arrangements also applies to humans. Extramarital sex certainly occurs, although not as frequently as media depictions or faulty statistics would have us believe. Various surveys of long-term partners, including a 2004 survey by biologist Leigh Simmons of four hundred students of both sexes, have revealed that one in four will engage in extrapair copulation at some time. The good news is that three-quarters of couples remain faithful over the long term. Moreover, the reported average rate of extrapair paternity is only around 2 percent—a conclusion confirmed by Alan Dixson in his 2009 book *Sexual Selection and the Origins of Human Mating Systems*. In other words, in humans a relatively low frequency of extrapair copulation results in an even lower level of extrapair paternity. Although some studies have reported a level of extrapair paternity as high as 12 percent, this seems to be exceptional. We can forget those apocryphal stories of "unpublishable" genetic studies in disadvantaged urban communities revealing that half the children had not been sired by their supposed fathers. In fact, humans in the societies surveyed seem to be more consistently monogamous than most bird species that have been studied.

Strangely, people cherish two incompatible ideas: that monogamy is the standard arrangement for human mating, and that men are less faithful than women. The original version of the doggerel quoted above, reportedly penned by Mrs. Amos Pinchot on awakening from a dream, was: "Hogamus, higamus, men are polygamous; higamus, hogamus, women monogamous." In a similar vein, American journalist H. L. Mencken once acidly remarked: "The only really happy folk are married women and single men." But this notion presents us with a problem: If women are typically monogamous, where do polygamous men find their extra partners? This enigma is highlighted by the finding, reported in many surveys, that men have more sexual partners than women. If, say, men report an average of ten sexual partners and women report only four, who are the six extra female partners reported by each man? One common explanation has been that, despite the promised anonymity of surveys, bravado drives men to overreport numbers of partners, while modesty drives women to underreport. Basic mathematics tells us that in a monogamous society there are only two possibilities: Either women and men are unfaithful in equal numbers, or a limited number of very promiscuous women must cater to the demands of many unfaithful men. As it turns out, a recent study revealed that prostitutes account for the inflated numbers of female partners. Apparently most men are too coy to admit that they paid for the extra experience.

Which brings us to a fundamental evolutionary question: Are humans biologically adapted for a particular pattern of social organization and mating? Cross-cultural evidence indicates that, as a species, we are highly variable in both respects. Evidence from comparisons with nonhuman primates is equally inconclusive. Despite this ambiguity, many authors resort to a shortcut. They simply take the chimpanzee as a frozen ancestor for human origins, concluding that our evolution began with promiscuous, multimale groups. At the other extreme, it is reasonable to argue that available evidence is so weak that we really cannot reach any firm conclusion regarding our ancestry. Indeed, many would conclude that there is no biological basis whatsoever for human social organization or mating arrangements and that social convention governs everything. According to this view, monogamous marriage is a purely social construct free of any biological predisposition. It turns out that both extreme views are untenable.

Differences in adult body size between males and females provide one important clue to social organization. In some primate species, males and females are almost equal in size (monomorphic), whereas in others they are significantly different (dimorphic), usually with males being bigger than females. Crucially, pair-living primate species are typically monomorphic. Males and females are approximately the same size, with an average difference below 15 percent. In harem-living and multimale species, by contrast, sexual dimorphism is common, although its extent varies widely. In extreme cases, such as African mandrills, males weigh more than twice as much as females. Humans are mildly sexually dimorphic. Worldwide averages indicate that men typically weigh just over 20 percent more than women. The actual degree of sexual dimorphism is really somewhat greater, because fat deposits account for a distinctly larger proportion of body weight in women. In prime adults, fat makes up about a quarter of body weight in women but only a tenth in men. This marked sex difference in body fat is unique among primates. Over and above this, men and women differ greatly in appearance because the fat is differently distributed. This sexual dimorphism in body size and shape provides a clue that humans are not biologically adapted to live in pair-based groups.

No discussion of human mating arrangements can gloss over the topic of incest. Incest is commonly defined as mating with close relatives, although a book review in *Nature* more pithily called it "multiplying without going forth." The key point is that mating between close relatives is likely to be detrimental because of inbreeding. All human societies have some kind of incest taboo, but the specific kinds of relatives excluded differ from society to society. Unions between parents and children or among siblings are usually prohibited as a matter of course, but uncles, aunts, and especially cousins may be viewed as acceptable partners depending on the culture. Regarding marriage between first cousins, for instance, the classical Greek and Roman worlds parted company. Athens and Sparta raised no obstacle, while Rome was vehemently opposed. As a Protestant, Charles Darwin was able to marry his first cousin Emma Wedgwood without the special dispensation traditionally required for Roman Catholics, although he later worried about the dangers of marriage between close relatives.

Eminent thinkers such as Sigmund Freud and Claude Lévi-Strauss fostered a myth that incest taboos are a purely cultural construct, unique to humans. They supposed that other animals mate indiscriminately, while humans uniquely benefit from a socially prescribed taboo. However, the notion of indiscriminate mating in other animals is simply wrong. Inbreeding increases the expression of otherwise rare genetic conditions that are often harmful, so we can confidently expect natural selection to favor mechanisms that curtail mating between closely related individuals. And it does. Among mammals, inbreeding is mainly avoided through the mechanism of individuals dispersing from their place of birth. This works best if only one sex migrates; if both males and females migrate, it is possible for related individuals to end up together again. The general rule for mammals is that males migrate and females stay put, matching the expectation that individuals of the sex that invests least in offspring should move out. One outcome is that females within a group are often related, forming its social backbone.

In mammals, including primates, females clearly invest most in offspring and generally do not leave their family group. Many night-active primates show this pattern, and so do various monkeys and apes, including macaques, plains baboons, and black-and-white colobus. Long-term field studies eventually revealed a reverse pattern in some monkeys and apes, including chimpanzees, red colobus monkeys, and spider monkeys, in which males stay put and females migrate. There is no convincing explanation for these exceptions, but the result is that males, not females, form the backbones of social groups. The net effect—the avoidance of inbreeding—is the same. With pair-living primates things work differently. Here, both males and females must leave the area of birth as adulthood approaches. Thus related individuals migrating out of one group might possibly meet up in another. This occurrence can be prevented if migrating siblings avoid one another or if males, for instance, are programmed to migrate farther than females. Both mechanisms probably operate in practice.

Interestingly, in human societies it is typical for women to marry out and for men to stay put; some regard female dispersal as a human cultural universal. This surely overstates the case, but ingenious genetic analyses have confirmed that men tended to stay home far more than women during the evolution of our species. This finding is important for two reasons. First, female dispersal should reduce inbreeding. Second, evidence from non-human primates indicates that a pronounced bias toward female migration is unlikely in a species biologically adapted for monogamy.

Avoidance of inbreeding is certainly not unique to humans. But explicit incest taboos that vary from one society to another are. Why do we need them? Undoubtedly, we descended from ancestors with natural mechanisms that precluded mating between close relatives. Is it at all likely that those mechanisms disappeared early in human evolution only to be replaced by incest taboos? Rather than replacing natural mechanisms for inbreeding avoidance, incest taboos serve to strengthen them. For instance, if early humans, for some reason, were less able to disperse to avoid inbreeding, perhaps an additional aid was needed to block mating between close relatives. One simple possibility is that males and females that grow up together find each other unattractive as mates. There is evidence for such a "kibbutz effect" in humans. As far as mating is concerned, familiarity does breed contempt. Social reinforcement of inbreeding avoidance may have been needed because social organization and mating arrangements became far more flexible in humans than in other primates. With that said, we can now return to exploration of the biological basis for human mating patterns.

In examining mating patterns, we have to consider the possibility that sperms from different males may have to compete. In this respect, primate groups containing a single adult male (pairs or harems) differ fundamentally from those with several adult males. In a one-male group, the resident adult male faces no direct competition when mating. In a multi-male group, by contrast, mating is typically promiscuous to some degree, making direct competition between males and their sperms likely. There is, of course, a caveat: If females in one-male groups engage in sneaky matings with extragroup males, then there will be some competition between males. For the sake of argument, however, let us accept the idea that direct competition between males is unimportant in single-male groups but prevalent and perhaps even fierce in multimale groups, where adult males try to exclude one another from mating with resident females. Often adult males form a relatively stable dominance hierarchy, established and maintained by competitive encounters. It is generally accepted that high-ranking males will have easier mating access to females. Nevertheless, two or more males may copulate with a female during a single ovarian cycle.

This brings us to the topic of sperm competition, which opens up new avenues of thought about whether humans are biologically adapted for a particular mating system. The underlying reasoning is quite simple. Active testes are expensive organs, using about as much energy as a chunk of brain of the same size. (Indeed, a feminist friend dismissively calls the testis the "male brain." This is not entirely without precedent. Leonardo da Vinci's famous drawing *The Copulation* shows an imaginary duct connecting a man's brain directly to his penis.) Because of their high energy demand, natural selection constrains testes to be just big enough to serve their function. When males live in multimale groups, natural selection should favor increased sperm production to boost the chances of mating success, and so males can be expected to have relatively large testes. By contrast, males living in one-male groups have little exposure to sperm competition, so their testes should be smaller. We can test these predictions with comparisons between species, although it is important to scale to body size. Other things being equal, males of larger-bodied species are likely to have bigger testes anyway, and this must be taken into account.

Zoologists Alan Dixson and Alexander Harcourt, among others, have conducted several broad-based studies of testis size in primates and other mammals. Their results have confirmed the predicted relationship of testis size with mating system. In monkeys and apes, for example, species living in multimale groups—such as macaques, baboons, and chimpanzees—all have large testes relative to body size. By contrast, males living in one-male groups typically have relatively small testes. This rule applies not only to pair-living species such as woolly lemurs, marmosets, owl monkeys, and gibbons but also to primates living in harem groups, such as various leaf monkeys and gorillas.

Human testes are relatively small. Men certainly have far smaller testes than chimpanzees, despite our larger body size. Whereas a human testis is about the size of a walnut, a chimpanzee testis is as big as a large chicken's egg. The small size of our testes directly conflicts with any suggestion that human social and mating systems are biologically adapted for chimpanzee-like promiscuity. Judged on size alone, human testes are seemingly adapted for a one-male mating system without sperm competition. Of course, this does not tell us whether men evolved to live in pairs, harems, or an orangutan-like dispersed system.

Several other dimensions of the reproductive tract support the evidence from relative testis size regarding biological adaptation for human mating, as Dixson demonstrates. For instance, the vas deferens—the duct that conveys sperms from the testis—is shorter and more muscular in primate species living in multimale mating groups, which are subject to sperm competition. In men, by contrast, the duct is quite long and only moderately muscular. Primates with large testes also have large seminal vesicles, indicating an enhanced capacity to produce seminal fluid. In humans, medium-sized vesicles produce about two-thirds of the seminal fluid, while the prostate gland adds the remaining third. Similar distinctions are found in female primates. For example, in species with multimale groups the oviducts are relatively long, increasing the distance that sperms must travel to fertilize the egg. Oviducts are contrastingly short in species living in one-male groups, and the relatively short oviducts of women clearly fall into this category. All evidence combined indicates that the reproductive systems of both men and women are adapted for a one-male mating context with little sperm competition.

Arguments based on testis size and other dimensions, such as seminal vesicle size or girth of the sperm-carrying duct or oviduct, may be flawed. The unstated assumption is that dimensions of the reproductive organs of any given species are genetically controlled and fixed within certain limits. Yet, as noted in the previous chapter, in seasonally breeding primate species testis size usually varies markedly across the year, so seasonal variation in testis size must be taken into account. It is also possible that—regardless of seasonal variation—testis size and other dimensions are adjusted to fit local conditions. This is suggested, for example, by minor variation in testis size between human populations. Reportedly, testes tend to be smaller in Asiatic men and larger in Europeans, particularly Scandinavians. It is not at all clear whether such variations in human testis size are genetically determined or influenced by social, nutritional, or other factors.

Alan Dixson, together with Matt Anderson, elegantly resolved the issue of environmental influence. In addition to examining overall testis size, which may be influenced by local conditions, they proceeded to examine sperms themselves. Remember that a sperm consists of a head containing the nucleus, a midpiece packed with mitochondria, and a whip-like propulsive tail. Mitochondria in the midpiece provide energy to power the tail as the sperm wriggles toward its goal. Dixson and Anderson reasoned that sperms exposed to competition might have a bigger midpiece, equivalent to a bigger fuel tank. To test this possibility they measured sperms of various primates. In this case, comparisons are straightforward because, remarkably, sperm size is not related to body size. Therefore scaling analyses are unnecessary. (Hold that thought.) Anderson and Dixson found a convincing relationship between sperm midpiece size and social organization. In multimale species such as macaques, plains baboons, and chimpanzees, the midpiece is significantly larger than in single-male species such as pair-living marmosets or gibbons and harem-living gelada baboons or gorillas. In human sperms, the midpiece is quite small, clearly falling into the range of primates living in one-male groups and well below values for species with multimale groups.

So results from measuring sperm midpiece size broadly match those obtained from comparative analyses of testis size. However, there are some important differences. For example, lesser mouse lemurs have relatively large testes, resembling the condition in primates living in multimale groups, whereas the sperm midpiece is quite small, falling into the range for primates living in one-male groups. Moreover, gorillas have notably small testes even compared to other primates living in one-male groups, whereas their sperm midpieces are among the largest for harem-living primates. It is clear that testis size and midpiece size can vary independently to some extent. Nevertheless, results for humans are quite explicit: Human males have relatively small testes and their sperms have an especially small midpiece, among the smallest recorded for any primate. There is no evidence whatsoever that human testes are biologically adapted for a mating context with pronounced sperm competition. Although environmental conditions may influence testis size, they do not affect sperm dimensions. Dimensions of sperms are quite constant in humans and are likely under tight genetic control. Therefore the sperm midpiece provides us with one of our strongest clues to biological adaptation for human mating.

Another piece of evidence that indicates the absence of sperm competition arises in the female. In some primates and other mammals, a plug forms in the female's vagina after mating. Rhesus monkeys provide one well-known example. Among apes, only chimpanzees produce a firm copulatory plug. Orangutans show persistent mild coagulation of semen but do not form a proper plug. One plausible explanation for copulatory plugs—graphically described as "competitive corking" by geneticist Steve Jones in *Y: The Descent of Men*—is that they block subsequent insemination by another male. Although human semen forms a dense mass almost immediately after ejaculation, it liquefies within fifteen minutes and does not constitute a plug. This finding indicates that humans are not biologically adapted for direct mating competition between males.

Substantial evidence from gene sequencing now supports the interpretation that primates with relatively large testes, such as rhesus monkeys and chimpanzees, are specially adapted to produce copulatory plugs. The main proteins in primate ejaculates, produced by the seminal vesicles, are two kinds of semenogelin, which are directly involved in coagulation. Geneticists Michael Jensen-Seaman and Wen-Hsiung Li examined the evolution of the two semenogelin genes found in both apes and humans and found that our genes have changed relatively little from the likely common ancestral condition. But in common chimpanzees the first of these two genes has almost doubled in length. In contrast, both semenogelin genes of gorillas actually show signs of degeneration. These findings suggest that the promiscuous mating of chimpanzees is a secondary condition that was not present in the common ancestor of gorillas, chimpanzees, and humans. That suggestion is supported by the fact that the length of the first semenogelin gene in bonobos has not increased as much as in common chimpanzees.

Later research examining rates of evolution of the second semenogelin gene has indicated that evolution occurred more rapidly in primates exposed to mating competition among males than in those living in one-male groups. Other studies have revealed strong positive selection on primate seminal proteins in general. One 2005 study comparing thousands of seminal genes between humans and chimpanzees identified, in addition to the two semenogelin genes, seven other genes with a strong signature of positive selection. The same study compared all nine genes in humans and chimpanzees with their counterparts in a dozen other primate species, including both Old World and New World monkeys. Strong signatures of positive selection were found in rhesus monkeys and baboons. Additional evidence of gene degeneration was found in gorillas and gibbons, both of which are characterized by single-male mating.

In sum, considerable evidence, ranging from testis size through sperm dimensions to genes controlling the composition of semen, indicates that humans are not biologically adapted to cope with high levels of sperm competition. On balance, it would seem that our species evolved to live in social groups with one-male breeding units, although there are some indications that mild competition between males might have existed at some stage of our evolution.

Apart from mating systems, sperm dimensions are also significant in an entirely different area. Recall that sperm size is unrelated to body size. This is also true of eggs and other body cells. The general rule is that bigger mammals have more cells, not larger ones. What is remarkable is that sperms and eggs are about the same size in a two-ounce lesser mouse lemur as their counterparts in a human 1,000 times greater in weight. More striking yet, sperms and eggs are about the same sizes in a blue whale weighing 3,000 times more than a human. This astonishing fact leads to a crucial but generally neglected problem.

A mammal egg is only barely visible, about as big as the period at the end of this sentence. Sperms are even smaller and invisible without a microscope. The volume of a mammal egg is equivalent to 30,000 sperms, and about 3 billion sperms would fit inside a peppercorn. The tiny size of sperms and eggs means that contriving an encounter between them is a considerably greater logistical problem for humans than for mouse lemurs. After insemination, human sperms have to travel about ten times farther than mouse lemur sperms. Moreover, they have to home in on an egg inside an oviduct that is ten times wider. Yet the midpiece of a mouse lemur sperm—its "fuel tank"—is actually somewhat larger than that of a human sperm.

For a blue whale, the logistics of egg-sperm encounters are awe-inspiring. Sperms have to travel almost twenty times farther than in a human and two hundred times farther than in a mouse lemur.

Even though there is a scaling problem with sperm dimensions, in this case body size is not an impediment to meaningful comparisons. Instead, it informs us about the mechanics of fertilization. Sperms and their midpieces do not increase in size as mammals become larger, so sperm transit up the female tract becomes more challenging as body size increases. Because sperms generally do not have a bigger fuel supply in large-bodied mammals, they need external assistance to travel toward the egg. The mechanics of copulation may provide some clues, but we are left with an inescapable conclusion: The female's womb and oviducts must actively assist sperm transit in some way, at least in large-bodied mammals. Women fall under this category, so we surely require mechanisms that promote sperm transport after insemination. When moving unaided, human sperms can swim about seven inches in an hour. At that rate it would take at least forty-five minutes for sperms to cover the distance from the neck of the womb to the lower end of the oviduct—assuming that they carry enough fuel for such a long swim. Of course, it would take even longer for sperms to travel to the oviduct from the site of natural insemination in the vagina. To give an idea of such a feat, the total distance covered by a human sperm from insemination to fertilization is equivalent to a four-mile, ninety-minute swim for a trained channel swimmer.

IN FACT, IT TURNS OUT that a woman's womb is pumping away unobtrusively. Various studies have used inert particles rather than actual sperms to test for transport mechanisms in women, canceling out any contribution made by the sperms' own movements. Early work on cows indicated that some sperms reached the oviduct less than three minutes after insemination. Following on from this and some incidental reports on humans, obstetricians Gene Egli and Michael Newton undertook a pioneering study in 1961. They conducted trials with three volunteers who had requested hysterectomies, timing the operations to take place close to the expected day of ovulation. In each case, carbon particles suspended in fluid were introduced deep in the vagina. In two of the three women, particles were recovered from the oviducts during surgery about half an hour after introduction. Thirty minutes was thus the maximum time taken for transit. Egli and Newton concluded that there must be some form of active transport, with muscular contractions of the womb probably playing an important part.

In a similar vein, in 1972 gynecologist Charles de Boer published a study on transit of particles through a woman's womb and oviducts. De Boer injected a small quantity of India ink at various sites in the vagina or womb in almost two hundred patients scheduled for surgery, either to remove the womb or to tie off the oviducts. Transfer of India ink from the vagina to the oviduct occurred in only 6 percent of patients. By contrast, transfer from the neck of the womb to the oviduct took place in almost a third, while transfer from the womb chamber itself occurred in over half. In some patients, substantial amounts of India ink spilled into the abdominal cavity after passing through the oviducts. In conclusion, de Boer suggested that sperms have to swim independently to make their way into the neck of the womb, but after that, muscular action of the womb and activity in the oviducts ensures transit toward the ovaries.

Gynecologist Georg Kunz took a different approach to studying sperm transport through a woman's tract. Kunz and his colleagues used a combination of two methods. In thirty-six women they recorded waves of muscular contraction in the womb, using ultrasound scans with a probe placed in the vagina. In sixty-four women they studied passive transport directly, by depositing sperm-sized albumin spheres close to the lower end of the neck of the womb. The spheres were radioactively labeled so that their passage to the oviducts could be tracked with a detector. Spheres reached the entrance to the oviduct as early as one minute after deposition, which meant the womb was clearly acting as a pump. Indeed, particularly around the time of ovulation, contraction waves were directed mainly from the neck of the womb to the oviduct entrances. Because a woman usually releases one egg in each cycle, only one oviduct—left or right—is needed for fertilization in each cycle. Interestingly, labeled spheres were mainly transported into the oviduct on the side of egg release, which matches findings from other studies. There clearly exists a mechanism to ensure that most sperms will arrive at the site where fertilization can occur.

Before we move on from the topic of sperm transit, one curious observation deserves special mention. In some women, one oviduct is blocked, often as a sequel to infection. Fertilization is not expected to occur when an egg is released from the ovary on the same side as the blocked oviduct. But in several cases fertilization nonetheless ensued. Unfortunately, this leads to implantation and development in the upper end of the blocked oviduct, making surgery inevitable. The only explanation for such an occurrence is that sperms migrated up the other, unblocked oviduct. To reach the egg discharged into the blocked oviduct, the fertilizing sperm must have swum around through the abdominal cavity, bearing eloquent witness both to the doggedness of sperms on their way to the egg and to their ability to swim the final stretch.

Having discussed the relationship between anatomy, the size of sex cells, and how social structure connects up with mating arrangements, we come to the inevitable question of how they intersect—which is through sexual intercourse. Popular books have established the notion that human copulation has two unique features: It can take place at any time in the ovarian cycle, and it occurs even during pregnancy. On this basis, Desmond Morris described humans as the "sexiest" primate species. Yet both claims to uniqueness clash with abundant biological evidence.

We can quickly dismiss the claim that copulation during pregnancy is unique to humans. It is easy to understand how this myth originated. Copulation after conception seemingly serves no useful biological function, so it is tempting to assume that it never happens with other mammals. But it most certainly does. It is widespread in the animal kingdom, as has been known for more than a century. A rapid survey at once reveals that copulation during pregnancy has been seen in mammals as diverse as shrews, mice, hamsters, rabbits, pigs, cows, horses, and primates, including tamarins, macaques, baboons, sacred langurs, and chimpanzees. Hormonal patterns associated with mating during pregnancy have been studied in rhesus monkeys. Copulation during pregnancy in other mammals mainly occurs early on and generally wanes as pregnancy progresses, but the behavior is so common that it surely must serve some function. Instead of heeding baseless claims that other mammals do not mate during pregnancy, we should seek explanations for why they do. Here is an enigma still awaiting solution. (Let it be noted, incidentally, that copulation is generally risk-free even during the latest stages of human pregnancy. Some studies have indicated that it may actually be beneficial in some respects.)

The claim that humans are the only mammals that engage in copulation throughout the ovarian cycle has a little more justification. However, our distinctiveness has been overstated and misrepresented. The crucial point is that copulation can occur at times during the cycle when ovulation is unlikely to occur: namely, we have sex whenever we feel like it. Such out-of-phase copulation is rare to nonexistent among mammals other than primates. Although absent among prosimians (lemurs, lorises, tarsiers), it is widespread among simians (monkeys and apes). Indeed, almost all simians copulate over a substantial part of the ovarian cycle. Gorillas are one of the few exceptions, copulating for only a few days in each cycle, close to the time of ovulation. As a general rule, monkeys and apes mate for at least a week during the ovarian cycle, and in some species copulation can occur on almost any cycle day. Overall, copulation typically occurs more often during the phase of follicular growth leading up to ovulation than in the luteal phase following ovulation, in which a corpus luteum, or yellow body, is present.

Most female mammals mate only during a tightly restricted period of one to three days in each ovary cycle. A female may actively solicit copulation during this period, and in common parlance is said to be "in heat" if she shows obvious readiness to mate during her receptive phase. Reproductive biologist Walter Heape coined the term "estrus" in 1900, referring to a restricted and obvious period of female readiness. The term's origin is quite curious—it comes from the Greek *oistros*, meaning "botfly." The family Oestridae contains about 150 botfly species. Larvae of these parasitic flies develop in tissue of a mammal host. Heape's connection between botflies and heat in female mammals was seemingly inspired by the fact that botflies can drive cattle into frenzy. Whatever his reason, the word "estrus" soon became the standard label for a limited period of mating readiness in female mammals.

Prosimian primates are like other mammals in that females typically have a limited period of estrus. By contrast, simians generally copulate over a large part of the ovarian cycle. Because of this, reproductive biologist Barry Keverne proposed in 1981 that we should avoid the term "estrus" for monkeys and apes. He was not merely quibbling over words. Substantial evidence indicates that in female simians direct hormonal control of mating readiness has decreased, while flexible control by the brain has increased. Unfortunately, Kerverne's well-founded arguments—which since have been supported by many other experts—are often overlooked. Misleading references to "estrus" in monkeys and apes abound. This oversight explains why a vital point has been widely ignored: Extended copulation during the cycle is both widespread among monkeys and apes and largely unique to them and humans. The finding is not new. Reproductive physiologist Sydney Asdell, author of the encyclopedic *Patterns of Mammalian Reproduction*, noted this unusual pattern for rhesus macaques more than eighty years ago. It is true that humans represent an extreme condition, with frequent copulation across the ovarian cycle. But extended mating over the female cycle was probably already present in the common ancestor of monkeys, apes, and humans more than 40 million years ago.

For mammals that copulate during a limited time in each cycle, ovulation, the release of an egg from the ovary, takes place within that brief mating window. Tightly confined copulation is precisely what we would expect to evolve, as it guarantees that a fresh sperm fertilizes a fresh egg. Mammal sperms generally survive for about two days after ejaculation, and eggs typically less than a day. Therefore it is difficult to understand why a female mammal would mate at any time other than close to ovulation. If copulation does not coincide closely with ovulation, a timeworn sperm might fertilize a freshly ovulated egg, or a decaying egg might be fertilized by a freshly ejaculated sperm. Early experiments on mammals such as rabbits and rats showed that fertilization with timeworn sperms or eggs can lead to pregnancy loss or birth of defective offspring. Why, then, do monkeys, apes, and humans copulate at times when fertilization with timeworn sex cells seems almost bound to happen? This crucial question, largely ignored, has profound medical implications, as we shall see in the final chapter.

Abundant evidence shows that ovulation occurs about halfway through the ovarian cycle in humans and many monkeys and apes, ranging from macaques, baboons, and langurs to chimpanzees. Thus any copulation a few days before or after midcycle could potentially cause problems. At best, copulation several days away from ovulation will simply be infertile. At worst, mating closer to ovulation, but not close enough, could lead to fertilization with a timeworn sperm or egg. Of course, monkeys, apes, and humans may have evolved a special mechanism to eliminate this potential problem. However, this assumption leads to another conundrum: Why would the ancestor of these simian primates lose an otherwise universal mechanism that ensures fertilization with fresh sex cells? If fertilization with timeworn sex cells leads to pregnancy loss and/or birth of defective offspring, strong selection should confine copulation to a brief window around ovulation. So how on earth could copulation outside the ovulation period ever evolve? Furthermore, there is another potential problem, which is that extended mating during the ovarian cycle, particularly if several males are involved, could increase any risk posed by germs borne on sperms.

Discussions of extended copulation during the ovarian cycle rarely mention possible dangers of fertilization with timeworn sex cells. Ignoring this issue, many scientists have proposed imaginative explanations for unrestricted mating. Anthropologist Nancy Burley linked extended copulation to "concealed ovulation." This idea rests on a common assumption that a male cannot detect whether a female is ovulating unless she emits a clear signal of some kind. Three main hypotheses, which are not necessarily mutually exclusive, have been put forth. The first is that copulation over an extended period in the cycle strengthens the bond between partners. Desmond Morris championed this proposal in *The Naked Ape*. Anthropologists Lee Benshoof and Randy Thornhill began an influential 1979 paper on the evolution of monogamy and concealed ovulation in humans with the following words: "*Homo sapiens* is unique among primates in that it is the only group-living species in which monogamy is the major mating system and the only species in which females do not reveal their ovulation by estrus." As a matter of fact, many primate species (approximately eighty) are monogamous, and ovulation is not revealed by estrus in most monkeys and apes. A second hypothesis proposes that extended female receptivity confuses paternity. If males cannot detect when ovulation occurs during a female's extended mating period, the paternity of offspring is not obvious. Such a mechanism could reduce competition between males, with various social benefits. The third proposal is that extended copulation might reinforce male investment in offspring. Nancy Burley in fact suggested a fourth explanation specifically tailored for humans: Ovulation time is "concealed" from women themselves, preventing deliberate avoidance of conception.

THE STANDARD "CLOCKWORK EGG TIMER" model of the ovarian cycle in monkeys, apes, and humans was heavily influenced by work on rhesus monkeys initiated in the 1920s by reproductive biologist Carl Hartman. Partly because of his studies, rhesus monkeys rapidly became the world's leading laboratory primates. Among other things, Hartman's research played a key part in establishing the idea of a "fertile period"—a clear mid-cycle peak of conceptions—in Old World monkeys. Closer examination, though, reveals that one of his interpretations, now taken as gospel by many, is fundamentally flawed.

Hartman's 1932 monograph on rhesus monkey breeding contains a diagram showing a peak of fertile matings between days nine and eighteen of the monthlong cycle. Note that fertile matings occurred throughout a ten-day interval—a third of cycle length—not just close to the middle, on day fourteen. But the real flaw lies elsewhere. Hartman's diagram was based on single occasions when a male and female were caged together. If Hartman had allowed his macaques to mate randomly across the cycle, his diagram would show when fertile mating was most likely. Instead, Hartman believed that mating leading to conception typically occurs at midcycle, so that is when he arranged most pairings. He arranged no pairings at all during the first and last weeks of the cycle. A midcycle peak was therefore a foregone conclusion. The real test of midcycle conception, using Hartman's data, is to calculate for each cycle day the proportion of mating opportunities that led to conception. After this is done, the apparent midcycle peak completely vanishes.

Almost forty years after Hartman's monograph appeared, a similar diagram for fertile matings in baboons was published, showing fertile matings concentrated on days nine to twenty of the cycle, a twelve-day spread. Again, most pairings were arranged at midcycle. When proportions of fertile matings are calculated, the midcycle peak vanishes, just as with Hartman's macaques. So two of the most widely cited examples of typical midcycle conception in Old World monkeys crumble under close scrutiny.

The notion of typical midcycle ovulation and conception in women was popularized by two gynecologists working in the 1920s and 1930s: Hermann Knaus in Austria and Kyusaku Ogino in Japan. Knaus and Ogino, perhaps more than anyone else, contributed to the emerging "egg timer" model: a monthlong cycle mechanically ticking away like a clock. Knaus was convinced that the ovarian cycle of rabbits provided a direct model for the human cycle. His choice was misguided. Ovulation in rabbits is induced, whereas women, like all other primates, ovulate spontaneously. Ogino, Knaus, and Hartman were contemporaries, and results emerging from studies of rhesus macaques also influenced interpretations of the ovarian cycle and conception in women. Thus Hartman's apparent evidence for midcycle conception in macaques fed into the standard model of the human ovarian cycle. Following Ogino and Knaus, many scientists reported evidence for a "fertile window" in the human menstrual cycle. While Ogino and Knaus both accepted the fact that women's menstrual cycles vary in length, they shared the mistaken belief that the interval between ovulation and the onset of the next menstruation, aka the luteal phase, is fixed at two weeks.

It is still generally accepted that ovulation and fertile copulation regularly occur at midcycle in humans. Yet a large body of little-cited evidence based on single copulations leading to conception reveals a marked spread across the human cycle. Such data—which are largely confined to medical publications in German—can be traced back at least as far as 1869, to a paper published by gynecologist Johann Ahlfeld. He cited some two hundred clinical records for cases reportedly based on single copulations leading to pregnancy. The records indicate that isolated copulations resulting in conception occurred on every single day of the cycle, although conceptions were far more frequent following copulation during the first half of the cycle. A peak occurred at around day six after the onset of menstruation, not on day fourteen. At least twenty later studies based on comparable information reported a similar pattern: Copulation leading to pregnancy occurred on virtually every day of the cycle but was far more likely during the first two weeks. Conception times were based on single copulations in various contexts, drawing on data for brief home visits by soldiers on military leave, legal deliberations in paternity suits, and records of gynecological clinics.

In the cases cited thus far, conception was not usually accompanied by physical violence. However, data for pregnancies arising as a consequence of rape support the view that copulation on virtually any day of the cycle can result in conception. To cite one example, in 1947 physician G. Linzenmeier examined 160 cases of women with regular monthly cycles who had been sexually assaulted. Sixty-two of them conceived, reportedly following rapes between day three and day eighteen of the cycle (a sixteen-day range). Linzenmeier concluded that this pattern was abnormal and that ovulation had been induced by the violence accompanying rape, an interpretation later adopted by veterinarian Wolfgang Jöchle. In fact, ovulation is typically spontaneous in primates and there is no convincing evidence that induced ovulation ever occurs in women. It is far more likely that a wide spread of copulation times leading to conception is a normal feature of the human cycle.

In my own recent analysis—which comprises data from ten studies that yielded a total of over 4,000 cases—I obtained a remarkably smooth curve leading to the following conclusion: Copulation resulting in conception can occur on almost any day of the cycle, although the probability of conception is higher in the first half, during the follicular phase, than in the second half, during the luteal phase. The curve shows a peak in the first half of the cycle, but it is closer to days eight or nine than to midcycle day fourteen.

The earlier reports supporting this conclusion have generally been ignored. Some scientists have dismissed them as unreliable because they depend on circumstantial evidence and were published before hormonal monitoring became readily available. The reports also rely on recollection by individual women regarding dates of menstruation and copulation, and such testimony is often said to be unreliable. Nevertheless, a 2000 paper by epidemiologist Allen Wilcox confirmed those earlier reports. His research group enlisted more than two hundred healthy women in North Carolina who had stopped birth control and were planning to become pregnant. Each subject recorded days on which intercourse and menstruation occurred. Ovulation times were reliably estimated with hormone assays of urine samples. Sixty percent of the women conceived and had live births. Peak probability of conception was found on days twelve and thirteen, but any woman had at least a one-in-ten chance of being in her fertile window between cycle days six and twenty-one. Conception outside that range was rare but still possible. Wilcox and colleagues specifically cited one of the earlier German studies and commented on the close correspondence between their results. They concluded: "Women should be advised that the timing of their fertile window can be highly unpredictable, even if their cycles are usually regular. . . . Our data suggest there are few days in the menstrual cycle during which some women are not potentially capable of becoming pregnant."

Gynecologist Yutaka Yoshida provided valuable information on the timing of ovulation and conception in a 1960 paper on artificial insemination using donor semen. All husbands in the couples involved were regarded as absolutely sterile because sperm counts were very low to zero. In contrast to many other practitioners, Yoshida carried out only one donor insemination in each cycle, eventually reporting on more than a hundred instances of conception resulting from a single insemination. Two indirect indicators of ovulation—basal body temperature and condition of mucus secreted by the neck of the womb—were used to optimize timing of insemination. Inferred ovulation occurred on days ten to twenty-three of individual cycles and Yoshida performed successful insemination on days eight to twenty-two, with a peak on day fourteen. Most interestingly, single inseminations leading to pregnancy covered an overall range extending from minus ten to plus four days relative to estimated ovulation time. This result provides evidence that human sperms may survive for up to ten days following insemination. That span is notably greater than the two-day survival for sperms that is generally accepted. Although Yoshida made a concerted effort to coordinate artificial insemination with ovulation, he was impeded by natural variation in menstrual cycles. Furthermore, when his results are analyzed in terms of the proportion of successful inseminations on any cycle day, there is no obvious peak. The same flat distribution is found as for single timed matings of rhesus monkeys and baboons.

THERE IS CLEARLY SOMETHING UNUSUAL about mating and conception in simian primates, as copulation leading to conception can evidently occur over a large part of the ovarian cycle in many monkeys and apes, as well as in humans. A fresh perspective on this distinctive feature came from a surprising source. In a key 1982 paper, zoologist Richard Kiltie surveyed pregnancy lengths of almost fifty mammal species and found that average variation was 3 percent overall. But his survey included five primate species—all simians, as it happened—whose pregnancy lengths were more variable. Intrigued by this, I compiled my own data. For twenty-seven nonprimates, standardized variation was generally close to 2 percent, and the same was true for twelve prosimians. By contrast, standardized variation for fifteen monkeys and apes was twice as high, averaging around 4 percent.

There are two potential explanations for this discrepancy. One is that pregnancies are simply more variable in simian primates than in other mammals. However, mammal pregnancies generally show unusually precise timing; why should this precision be relaxed in monkeys and apes? The second possibility is that greater variability in simian pregnancy length may be an artifact arising from the way in which conception is inferred. Pregnancy durations reported for simians come mainly from observations of copulation in captivity. In mammals with a clearly defined estrus, including prosimian primates, the conception date can generally be pinpointed quite accurately by observing when mating occurs. For most mammals, variation in pregnancy lengths determined from mating generally reflects real variation in the interval between conception and birth. Monkeys and apes are quite different. Copulation often occurs on several days of the ovarian cycle, so mating does not precisely indicate conception time. It is widely taken for granted that ovulation and copulation leading to conception both typically occur at midcycle. But if conception can result from copulation at other times, this could explain the greater variation of calculated pregnancy lengths. If sperm can be stored somewhere in the female reproductive tract, for instance, that could explain a gap of several days between mating and the time of conception. This hypothesis leads directly to a testable prediction: Variability of pregnancy lengths recorded for monkeys and apes should decrease when conception time is estimated with more reliable evidence.

One simple way to obtain more reliable information is to restrict encounters between males and females. In various studies of rhesus macaques, for instance, females had different degrees of exposure to males. With extended access, pregnancy lengths were reported to have standardized variation of almost 5 percent, over twice the level found in most mammals. When exposure of the female to the male was limited to less than fifteen minutes on a single day, variation declined to about 3 percent. That said, if sperm are stored somewhere in the female tract, even a precise mating time will not reliably indicate conception time.

Instead of relying on observations of copulation, researchers increasingly use hormonal monitoring of female cycles to pinpoint actual ovulation time. Availability of sensitive methods for measuring hormone levels in urine or feces has made this much easier. Ovarian cycles and pregnancy can be monitored in a "hands-off" manner, without repeated stressful captures of females to take blood samples. I first became involved in such work in the 1970s at the London Zoological Society, working with reproductive biologist Brian Seaton on urine samples from gorillas. Thereafter, I and other colleagues monitored urinary hormones in various other primate species. For a project on Bolivian Goeldi's monkeys led by reproductive biologist Christopher Pryce in Zürich, we needed an accurate figure for the normal pregnancy length. Previous values based on indirect evidence, notably copulation, were conflicting and inconclusive. Goeldi's monkeys mate infrequently and irregularly across the ovarian cycle and also mate during pregnancy, so copulation is an unreliable indicator of conception time. Hormonal data permitted us to pinpoint conception within a three-day period, yielding an average pregnancy length of 152 days. Variation was only 2 percent, so accurate measurement revealed that pregnancy duration in Goeldi's monkeys is actually no more variable than in mammals with a clearly defined estrus.

Humans, on the other hand, are a challenging case when it comes to determining the duration of pregnancy. Women give no easily detected sign of ovulation time. In medical circles, the duration of pregnancy is usually calculated from the first day of the last menstruation before conception. As a result, the commonly cited human pregnancy length of forty weeks, just over nine months, is not comparable with pregnancy durations determined for other mammals, which are all dated from the time of conception. Taking the common assumption that conception typically occurs on day fourteen of the human cycle, the true average for human pregnancy length should be about thirty-eight weeks. A 1950 paper by physicians J. R. Gibson and Thomas McKeown reviewed more than 17,000 pregnancy durations in a complete annual sample from women in Birmingham, England, for 1947. They considered only cases leading to live birth of single infants. Pregnancy lengths calculated from the last menstruation yield an average value a shade above 280 days, with standardized variation of just over 5.5 percent. Despite the huge sample size, this standardized variation is one of the largest recorded for simian primates and hence for mammals generally.

As with nonhuman primates, a more precise definition of likely conception time yields less-variable estimates of human pregnancy length. For example, in R. Dyroff's 1939 study of human pregnancies resulting from brief visits by male partners on military leave, the coefficient of variation was reduced to about 4.5 percent. In a later study, the coefficient of variation was reduced even further, to about 3.5 percent. Incidentally, these reductions in estimated pregnancy length indicate that women's testimony for those studies was actually quite reliable.

Procedures used to treat infertility provide even more accurate information on human pregnancy length. Over past decades, millions of women have conceived as a result of interventions such as artificial insemination and IVF. Here is a golden opportunity to determine the length of human pregnancy more precisely. Unfortunately, this question has not been a priority for fertility specialists, and vast quantities of raw data have remained unpublished. Happily, while I was writing this book, I was introduced to Howard Hamilton, the executive director of Fertility Centers of Illinois. This organization is by far the largest devoted to infertility treatment in Illinois, and a guided tour through its state-of-the-art facility in Chicago simply bowled me over.

Dr. Hamilton, as well as resident fertility doctors Kevin Lederer and Aaron Lifchez, were very receptive to my aim of analyzing data to test the possibility of sperm storage. Indeed, Dr. Lifchez noted a key point that I had overlooked. Nowadays, fertility specialists prefer to carry out artificial insemination by injecting semen directly into the womb in a process called intrauterine insemination, as this increases the success rate. Dr. Lifchez astutely pointed out that intrauterine insemination bypasses the neck of the womb, so if that is the site of sperm storage, there should be no delay between insemination and conception. My expectation was that analysis of data for intrauterine insemination and IVF would yield a more precise figure for the real duration of human pregnancy, from conception to birth. Having taken all necessary measures to protect patient confidentiality, I was allowed to process a large data set from Fertility Centers of Illinois, aided by my intern Hannah Koch.

Almost at once we hit a snag. Premature birth is significantly more common following intrauterine insemination and IVF, perhaps because these procedures bypass the natural filtering action of the neck of the womb. Intervals between conception and birth are more variable as a result, so we could not use variation in pregnancy length to indicate precision. Fortunately, it is possible to sidestep this issue by examining the relationship between birth weight and the interval between insemination or fertilization and birth. In principle, the weight of the newborn should increase with the time it spends in the womb. However, birth weight is only weakly related to pregnancy length dated from the first day of the last menstruation. This is one problem that arises when the time of conception is not reliably known. With IVF, by contrast, conception time can be dated exactly. Sure enough, when we analyzed data for some three hundred cases, the relationship between birth weight and IVF-to-birth interval turned out to be much tighter and highly significant. The trend revealed that birth weight is closely linked to time spent in the womb once the conception date is accurately known. We can now address the issue raised by Dr. Lifchez. Analysis of almost a hundred cases revealed that the relationship between birth weight and interval from intrauterine insemination to birth is just as tight as with IVF. This suggests that sperm are indeed stored in the neck of the womb and that insemination directly into the womb is quite rapidly followed by conception.

Part of the reason human conception times are so difficult to determine is because there is no externally visible sign of ovulation time. Many authors have claimed that "concealed ovulation" is unique to our species. While chimpanzees and various Old World monkeys do show prominent, externally visible sexual swellings that roughly indicate ovulation time, most primates lack any external physical sign of ovulation. All prosimians lack sexual swellings, as do all New World monkeys. Chimpanzees alone among the apes show prominent sexual swellings. In fact, sexual swellings are so uncommon that zoo visitors often misinterpret them. Zoo curators in both London and Zürich showed me irate letters from visitors insisting on surgical removal of unsightly growths from the bottoms of female chimpanzees. Yet such swellings are entirely natural; male chimpanzees presumably think they are cute.

The fact of the matter is that most simian primate species are like humans, lacking any externally visible sign of ovulation. Moreover, accumulating evidence reveals that even in simians with apparent external indicators of ovulation, the signals are unreliable. In prosimians, ovulation time is at least indirectly evident, as they have a tightly defined estrus like other mammals. In simians, by contrast, concealed ovulation is the rule, not the exception. Thus concealed ovulation is probably a feature that originated in ancestral simians, along with extended copulation during the ovarian cycle and greater variability in intervals between copulation and birth. The really interesting question, then, is why some Old World monkeys and chimpanzees have obvious sexual swellings that do provide an approximate external signal of the time of ovulation. But that is another story.

One special aspect of human copulatory behavior remains to be discussed, as it links up with the timing of conception during the cycle. This is variation in sexual motivation and activity across the ovarian cycle in women.

MUCH HAS BEEN WRITTEN ABOUT the purportedly unique "loss of estrus" in women. At the same time, there has also been great interest in the notion that at least some form of cyclical variation in sexual motivation has persisted. While it is true that copulation can take place at any time in humans, there is a clear pattern of hormonal change across the cycle. This pattern is similar to that in prosimian primates and various other mammals with a distinct estrus. For various reasons, millions of women now take natural or synthetic steroid hormones on a regular basis, but too little is known about their possible effects on sexuality, including behavior. In order to discern cyclical changes to sexual behavior, we need to understand any changes that accompany the hormonal pattern of the human cycle.

In 1933, anatomist Georgios Papanicolaou published a fascinating but long-neglected study of changes in the human vagina across the cycle. Sixteen years earlier, together with fellow anatomist Charles Stockard, he had invented a new method for studying the ovarian cycle of female hamsters, which required examining the cells in a fluid sample from the vagina. The fluid usually has a slimy consistency and includes cells shed by the vaginal lining along with red and white blood cells and abundant bacteria. A vaginal smear, as the preparation is called, reliably indicates ovulation time in various mammals. Papanicolaou later became famous for a related but different discovery, that a sample of cells from the neck of the womb, known as a cervical smear, permits early detection of cancer. Together with Herbert Traut, in 1943 he published *Diagnosis of Uterine Cancer by the Vaginal Smear*, thus originating the "Pap smear," now routinely used worldwide.

While his name lives on in his test, Papanicolaou's informative 1933 paper on monitoring the human cycle with vaginal smears drifted into oblivion. Over the past thirty years, apart from a few special cases, it has become standard practice to monitor cycles with hormone assays rather than vaginal smears. However, his findings continue to be relevant. Changes in vaginal smears across the ovarian cycle are more variable in women than in many other mammals, but a basic pattern can nonetheless be recognized. Taking results from more than a thousand individual smears, Papanicolaou divided the human female cycle into four basic stages: an initial phase, from the first day of menstruation to the seventh cycle day; a "copulative phase," from day eight to day twelve; a proliferative phase, extending up to day seventeen; and a premenstrual phase, between day eighteen and the onset of the next menstruation. Vaginal smears during the copulative phase show a pattern matching that seen during the mating period of rodents, including characteristic flat cells shed by the vaginal lining along with abundant mucus.

Papanicolaou specifically linked this copulative phase, from days eight through twelve, to reports that fertile copulation is prevalent during this period. He noted certain reports of a peak in desire during the period after menstruation, corresponding to his copulative phase. However, there were many conflicting accounts about when the peak occurred—other authors reported a peak just before menstruation, while yet others reported peaks both before and after menstruation. For instance, two peaks were suggested by Marie Stopes, the pioneering Scottish campaigner for women's rights who opened the first British birth control clinic in 1921. Her influential 1918 book *Married Love* (banned in the United States until 1931) included a diagram indicating that natural desire in healthy women peaks two to three days before menstruation and then again from eight to nine days after its end.

A major milestone regarding this topic was the publication of a 1937 paper by scientists Robert McCance and Elsie Widdowson. They used standard questionnaires to explore the relationship between hormones and psychology across the menstrual cycle. Unsurprisingly, when conducted seventy-five years ago, this survey encountered opposition. The authors report that some medical school heads "took exception to forms on the grounds that sexual feeling was abnormal in unmarried women students and that no forms containing such words could possibly be allowed to circulate in their institutions." Despite this, it eventually proved possible to review data from almost eight hundred complete cycles of almost two hundred women, more than half of them single. Married women in the study reported an average of five coital acts per cycle, with a maximum of eighteen. One striking finding was that copulation frequency peaked on day eight of the cycle, in the middle of the follicular phase. Contrary to expectation, then, the peak did not coincide with the most likely time of ovulation, on day fourteen. McCance and Widdowson prophetically commented on this mismatch: "If . . . the human sperm can rarely live in the female generative tract for more than two days, it is difficult to see why seven days should elapse between the period of maximum desire in the female and her time of ovulation."

FOLLOWING A HIATUS lasting more than twenty years, the question of cyclical variation in women's sexual desire resurfaced as a popular subject of scientific study. Particularly notable is a series of publications by public health scientists Richard Udry and Naomi Morris. After numerous analyses of data, they were able to discern a peak during the follicular phase. In 1971, biologist William James reanalyzed data from the two studies by McCance and Widdowson and by Udry and Morris, confirming their results.

Psychiatrists Diana Sanders and John Bancroft provided a useful survey of the relationship between female sexuality and the cycle in a 1982 paper. Their review included a summary of their own study of more than fifty women, carefully assessing mood and sexuality from daily diaries. In order to identify cycle phases, hormones were monitored in blood samples collected every other day or so. Sexual activity with the partner showed a cyclical pattern, with a significant peak in the middle of the follicular phase (days six through ten). A combined measure of sexual feelings revealed a similar peak in the middle of the follicular phase, but a second peak occurred during the late luteal phase, before menstruation. Some women indicated whether they or their partners initiated sexual activity. Initiation by the woman or mutually by both partners was also more likely to occur in the middle of the follicular phase. By contrast, initiation by the male partner tended to be more frequent during the luteal phase. Overall, this study provided evidence of an increase in sexual interest and activity in the middle of the follicular phase. There was also some evidence of an increase in sexual interest late in the luteal phase, although there was no indication of a midcycle peak around the time of ovulation.

Psychologist Harold Stanislaw and biologist Frank Rice took a different approach in their study of the relationship between sexual desire and the cycle. They noted that previous studies had often examined the distribution of intercourse across the cycle. Although it has often been reported that copulation is most frequent immediately after the end of menstruation, this could be a consequence of most couples abstaining from intercourse during menstruation. Frequency of copulation can also be influenced by factors such as day of the week, presence of a willing partner, male-initiated sexual activity, and avoidance of midcycle copulation as a contraceptive measure. All of these factors can reduce the association between copulation frequency and cyclical changes in female hormones. Stanislaw and Rice therefore designed a forward-looking study in which sexual desire could be recorded in relation to the rise in basal body temperature that often indicates ovulation. They found that, in any given cycle, sexual desire was usually first experienced a few days before the likely ovulation date.

Studies of the relationship between sexual desire and the cycle in women have yielded variable results. This is to be expected, as there are several confounding factors. In most cases ovulation time was not determined directly but only approximately inferred. Nevertheless, there are some general trends, which are modified if a woman uses oral contraceptives. If no steroid hormones are taken, female sexual motivation commonly shows a peak in the week after the end of menstruation (days eight through fourteen), sometimes followed by a secondary peak in the week preceding the next menstruation (days twenty-two through twenty-eight in a four-week cycle). Although sexual interest shows trends across the cycle, they are not obviously linked to circulating hormones. The premenstrual peak in sexual motivation has yet to be explained. However, the mid-follicular-phase peak occurs at about the same time as the copulative phase identified by Papanicolaou with vaginal smears. Moreover, as we have seen, abundant evidence from single occurrences indicates that fertile copulations also peak in the week following menstruation. Coincidence? I don't think so.

CHAPTER 4

Long Pregnancies and Difficult Births

In 2006, the permanent exhibit *Evolving Planet* opened at The Field Museum in Chicago. Ever since, I have seized every opportunity to stroll through the history of life on Earth. One of my favorite fossils on display is an exquisitely preserved ray from 50-million-year-old lake deposits in Wyoming. Skilled preparation exposed not only the skeleton of an adult female fish but also that of a baby developing inside her body. Rays and sharks count among a minority of fish that give birth instead of laying eggs. I also vividly remember my first encounter long ago with a fossilized pregnant ichthyosaur at the Natural History Museum in London. Ichthyosaurs, dolphin-like reptiles that lived in the seas between 250 million and 90 million years ago, disappeared some 25 million years before the dinosaurs went extinct. Many pregnant ichthyosaurs have been discovered. These are just two examples of animals with backbones, other than mammals, that have live births. In fact, more than a hundred modern nonmammal species have pregnancies, including various fish, a few amphibians (such as the aptly named midwife toad), and quite a few lizards and snakes among the reptiles. It never became the dominant lifestyle among nonmammals, though, and live births never evolved at all among birds. Given the high costs of pregnancy in energy and labor, how did this mode of giving birth evolve?

Mammal offspring have a huge head start in life because their mothers provide them with nourishment and a warm place to grow. The only exceptions are a few Australasian monotremes (platypuses and spiny anteaters), which cling to the ancient egg-laying habit. In the vast majority of mammals—namely, marsupials and placentals—the fertilized egg remains in the mother's womb and develops there. Live birth among mammals must thus date back over 125 million years to the common ancestor of marsupials and placentals, and perhaps even earlier. Although they left no surviving relatives, rodent-like multituberculates, with their peculiar many-cusped molar teeth, inhabited the planet for more than 100 million years, until they died out more than 40 million years ago. Their tiny pelvis was fused, leading paleontologist Zofia Kielan-Jaworowska to conclude that they could not lay eggs with enough yolk to develop outside the mother, and therefore gave birth to small live offspring instead. In other words, the beginnings of human pregnancy may date back more than 140 million years.

Human pregnancy, then, has a long evolutionary history. As with other aspects of human reproduction, exploring that history highlights and sometimes answers key questions. For instance, our immune system evolved to recognize and eliminate foreign proteins in our bodies. In that case, why doesn't a pregnant woman's body reject an embryo in her womb, an alien invader with many proteins that differ from her own? This question is especially apt because the human placenta is extremely invasive and there are few barriers to stop foreign proteins passing from the embryo to the mother's bloodstream. Related to this issue, deep invasion of the wall of the womb by the human placenta is generally seen as an adaptation for more efficient exchange between mother and embryo, one that has been linked to development of our very large brain. But is that explanation convincing? This is a prime case where comparisons with other primates and across all mammals can yield valuable clues. There are also basic questions about the length of human pregnancy. As discussed in the previous chapter, ovulation is not easily detected, so physicians have traditionally dated human pregnancy from the beginning of the last menstruation. Such a method permits only a rough prediction of when birth will happen. Can we improve on that? Finally, an understanding of evolutionary history also illuminates the origins of such curious topics as morning sickness, time of birth, and the mammalian practice of eating the placenta.

LIVE BIRTH HAS two huge benefits. First, developing offspring benefit from the protection of being inside a secure, temperature-controlled chamber. Second, resources transfer from mother to offspring with great efficiency. By comparison, laying eggs in a nest is a tremendous drain on resources and energy. Even birds, which are generally assiduous parents, risk exposing the eggs they incubate to cooling or predation if the nest is disturbed. It is also less efficient to convert maternal resources into a large yolk that must be reprocessed to produce an offspring after the egg has been laid. If live birth has so many obvious advantages, why do most other animals with backbones lay eggs (ovipary) rather than having live births (vivipary) like marsupials and placentals? Natural selection can only filter characters that arise through random mutation and recombination, so a chance development must have led to live birth in early mammals and certain advantages then led to its fixation as a universal feature of marsupials and placentals.

However, natural selection must also overcome several obstacles for live birth to become the dominant pattern. In mammals, once the embryo becomes intimately connected with the inner lining of the womb, compatibility issues between mother and offspring arise. Half of the offspring's genes come from its father, so it will produce many proteins unlike those of the mother. Intimate connection between the mother and her developing offspring makes nutrient transfer more efficient, but it also increases the challenge to her immune system. As the natural response is for the mother's body to attack and eliminate foreign proteins, special adaptations are needed to stop her immune system from rejecting the developing offspring.

Of course, in addition to having live births, marsupials and placentals also suckle their offspring. As egg-laying monotremes suckle, too, live birth evidently evolved later. Perhaps live birth followed by suckling, which is unique in the animal kingdom, is a winning combination. It also imposes a unique burden on females, as they are responsible for both pregnancy and suckling, with little direct contribution from males.

* * *

Before we can learn more about why live birth evolved, however, we must first examine the history of the womb, where it all begins. Fish, amphibians, reptiles, and birds generally do not have a womb, just oviducts to convey eggs to the outside world. Only mammals have a womb. In egg-laying monotremes, the lower end of each oviduct is expanded to form a structure sometimes called a womb. Yet a clearly developed womb is present only in marsupials and placentals.

Like most other bodily systems, the reproductive organs of any female mammal develop as right and left sides that are more or less mirror images. On each side of the body, a duct leading from the ovary to the exterior develops into a female reproductive tract with an oviduct, a womb, and a vagina. At first, left and right tracts develop quite separately, but later on they can merge to varying degrees. In all placental mammals, for example, left and right vaginal tubes unite to form a single vagina in the body's mid-line. Marsupials are different: They keep separate left and right vaginal tubes, and a special birth canal runs between them.

Apart from the vagina, mammals commonly retain two largely separate halves in the rest of the female reproductive tract. Marsupials and most placentals have left and right womb chambers, each connected with an individual oviduct. A double, "two-horned" womb of this kind exists even in the largest placental mammals, such as rhinoceroses, elephants, and whales. But a few placental mammals show an exceptional arrangement: Instead of separate left and right chambers, they have a single midline womb. Higher primates (monkeys, apes, and humans) have a single-chambered womb, as do sloths, armadillos, and a few bats. By contrast, lower primates (lemurs, lorises, and tarsiers) have all remained primitive, retaining two womb chambers like most other mammals. Therefore the single-chambered womb of higher primates is evidently a novel evolutionary development, dating back to their common ancestor around 40 million years ago.

Women normally have a single-chambered womb, reflecting their descent from the common ancestor of higher primates. Understanding the evolutionary background is medically important because developmental accidents do occur. During development, the human reproductive tract follows the ancient mammalian pattern at first. Harking back to an earlier evolutionary stage, it sometimes happens that left and right halves fail to fuse together. A two-chambered womb is a rare condition, found in 1 out of 3,000 adult women. In such cases, the vagina usually remains double as well. Remarkably, relatively normal pregnancy can occur in women with a two-chambered womb, including occasional multiple births, although complications such as premature birth are more frequent.

A single-chambered womb is rare among mammals, so its evolution must have involved some special selection pressure. It is likely related to the fact that higher primates are adapted for single offspring. For instance, as Chapter 6 will discuss, they all have only a single pair of teats, and each mammal species typically has one pair for each offspring born. Looking outside the primates, sloths have single offspring as well. At first sight armadillos seem to break the pattern, for they give birth to multiple offspring despite having a single-chambered womb. Yet there is an unusual twist to the story here. All offspring in an armadillo litter are clones derived by repeated division from a single fertilized egg. This means the common ancestor of armadillos was doubtless adapted for single offspring. The unique cloning pattern presumably resulted from later selection favoring multiple offspring to increase the rate of breeding. Ancestral adaptation for a single offspring might therefore provide part of the reason for development of a single-chambered womb, but this cannot be the whole answer. Although some lower primates have two to four offspring at each birth, most have singletons and some have only one pair of teats. In addition to lower primates, many other mammals typically have single offspring and one pair of teats and yet commonly retain a two-chambered womb.

To accurately trace the evolutionary pathway leading to a single-chambered womb, we need to see the intermediate stages. Although there are no halfway stages in primates, such intermediates exist in certain bats. The first important point is that all bats with a single-chambered womb, or some intermediate stage derived from the original two-chambered condition, have single offspring and one pair of teats. There, the connection between a single-chambered womb and single offspring holds up. Different bat species show various intermediate conditions. These range between mild enlargement of one womb chamber at the expense of the other and clear dominance of one chamber. This variation suggests, incidentally, that evolution of the single-chambered womb in simian primates, sloths, and armadillos might have occurred through enlargement of one chamber at the expense of the other, rather than through midline fusion of the two original chambers. However, we are still left with an unanswered question: Why do many mammals with single offspring and one pair of teats still have a two-chambered womb?

PREGNANCY IN MAMMALS starts with conception, when a single sperm fertilizes an egg. Fertilization takes place high up in the oviduct, in a slightly bulging region known as the ampulla. The early embryo then migrates down the oviduct toward the womb, repeatedly dividing during its ten-day journey. By the time it reaches the womb chamber, it has developed into a hollow ball of about a hundred cells, known as a blastocyst. Despite such rapid division, its overall size remains unchanged because the mother provides no nutrients until the blastocyst attaches to the inner wall of the womb in a process known as implantation. The outer layer of the blastocyst, the trophoblast, makes the initial attachment and may start to invade the inner lining of the womb. This is the beginning of the placenta. In mammals generally, including most primates, blastocyst attachment is superficial. In great apes and humans, by contrast, the blastocyst actually burrows into the wall of the womb, ending up enclosed in a small cavity. Such interstitial implantation is rare among mammals. Scientists do not really know why this happens, but it may provide the blastocyst with additional protection and speed up the development of the placenta.

Implantation opens up a direct lifeline between mother and offspring. It paves the way for efficient, direct nutrient supply and prompt elimination of waste products. Here, two basic terms referring to the offspring's development in the womb need clarification. The words "embryo" and "fetus" are often used interchangeably. For reproductive biologists, however, they have distinct meanings. The embryonic stage is the initial period during which different tissues are developed and the basic framework of the offspring's body is slowly formed. It begins with conception, passes through implantation, and then continues on through the initial period of exchange via the placenta. The term "fetus" is used once individual major organ systems of the body—brain, heart, gut, and urogenital apparatus—can be recognized, and from that point up until birth. In contrast to an embryo, a fetus resembles a newborn individual in broad outlines and differs mainly in its small size. In human development, the embryonic stage lasts eight weeks after conception, while the fetal stage takes up the remaining thirty weeks until birth.

* * *

WE CAN NOW GO ON to consider the topic of pregnancy duration, starting with an anecdote from anthropologist Mary Leakey. In her book *Disclosing the Past*, Leakey recounted her discovery of a partial skeleton of the 20-million-year-old fossil ape *Proconsul* on Rusinga Island in Kenya on October 2, 1948. Leakey rightly recognized this as a "wildly exciting find which would delight human paleontologists all over the world." To this day, *Proconsul* remains a key witness to the early evolution of the group containing apes and humans. This major discovery had another significant outcome as well: "When the magnitude of our discovery had sunk in, back in our camp at Kathwanga, Louis and I wanted to celebrate. We were exhilarated and also utterly content with each other and we thought that quite the best celebration would be to have another baby." Thus it was that their son Philip was born on June 21, 1949, after an interval of 262 days. Here, it seems, we have one fairly secure record of the human gestation period. If nothing else, it is a graphic example of how exploring our origins can contribute to human reproduction.

The duration of pregnancy, or gestation period, in any mammal is the interval from conception to birth. As discussed in the previous chapter, gestation periods are tightly controlled in mammals generally. Roughly speaking, standardized variation is plus or minus 2 percent. What does this mean for a human gestation period averaging about thirty-eight weeks, or 266 days, from conception to birth? It tells us that two-thirds of births should normally occur within a range of five days on either side of the average. At 262 days, Mary Leakey's *Proconsul*-triggered pregnancy fits within that range of 261–271 days. But one in twenty women would be expected to give birth more than two weeks before or after the thirty-eight-week marker. Although most normal human pregnancy lengths range between thirty-six and forty weeks, this is remarkably precise timing for a biological marathon with a multitude of complex processes. Be that as it may, prediction of a birth date with a range of plus or minus two weeks is still somewhat vague. How can a busy woman today be expected to plan properly using such imprecise information?

In fact, matters are considerably worse in practice, given the standard medical practice of timing human pregnancy from the first day of the last menstruation before conception, not from the date of conception itself.

Since 1812, physicians have generally used Nägele's rule to calculate the expected date of birth: The date is estimated by adding one year to the first day of a woman's last menstrual period, subtracting three months, and then adding seven days. This corresponds to a pregnancy duration of about forty weeks. On average, ovulation occurs about two weeks after the first day of the last menstruation. The medical definition, then, increases the actual gestation length of about thirty-eight weeks by that amount. Taking the standardized variation of plus or minus 5.5 percent found by the Gibson and McKeown Birmingham study discussed in the previous chapter, we would expect that two-thirds of human births predicted from the last menstruation should normally occur within a range of two weeks, instead of five days, on either side of the forty-week average. Ninety-five percent of births will lie within a range of four weeks on either side, between thirty-six and forty-four weeks. Still, in one out of twenty women normal birth will occur more than a month before or after the date predicted using an average of forty weeks. If we can't do better than that, pregnant women will continue to be forced to cope with a period of uncertainty lasting several weeks.

There is yet another reason for wide variation in human pregnancy lengths. Most women have one baby at a time, and singleton birth has been implicitly assumed in the discussion thus far. When more than one baby is born at a time, as occurs with more than 100,000 births in the United States every year, everything is radically different. The reason is quite straightforward: Because the womb cannot expand indefinitely, as the number of developing babies increases they are born smaller and smaller after ever shorter pregnancies. Multiple births also become rarer as the number of babies increases. In North America, twinning occurs about once in 83 pregnancies, and triplets once in 8,000 pregnancies. These figures fit the general rule that twins are not extraordinary, while triplets are quite rare, and rates decline even faster through quadruplets, quintuplets, sextuplets, septuplets, octuplets, and nonuplets. Just a few cases of nonuplets have been reported, but most of the infants were stillborn and none survived beyond a few days. Multiple human births with more than nine infants have never been reported.

In 1895, German physician Dionys Hellin noted an intriguing regularity in rates of human multiple births. If twins occur at a rate of, say, 1 in 80 births, triplets occur at a rate of 1 in 80^2 (1 in 6,400), quadruplets occur at a rate of 1 in 80^3 (1 in 512,000), and so on. Following this pattern, the rate of nonuplets should be 1 in 80^8, yielding the astronomically high figure of

1 in 1,677,721,600,000,000. This multiplication series is exactly what we should expect if there is a standard probability (1 in 80) of having an additional fetus in the womb. Although it has become known as Hellin's law, it is really a rule of thumb. Rates can also change over time: The older a woman, the more likely she is to have a multiple birth naturally. Moreover, rates of multiple birth differ between regions and may change over time within a single region. In 2011, Dutch economist Jeroen Smits and sociologist Christiaan Monden published a comprehensive survey of twinning rates across the developing world. They confirmed that twinning rates are low in East Asia and showed that this is actually the dominant pattern throughout the entire South and Southeast Asian region, with an average occurrence of twins in 1 of every 130 births. Comparable twinning rates are found in nations throughout Latin America. By contrast, the oft-mentioned, strikingly high rate originally reported for Nigeria was found to be typical for Africa generally, with an overall average of twins in 1 of every 60 births across the continent. The twinning champion of the world turned out to be Nigeria's neighbor Benin, with twins in 1 of 35 births. In comparison to Africa on one hand and Asia and Latin America on the other, Europe and North America have intermediate twinning rates, with twins in roughly 1 of 80 births.

A LANDMARK PAPER PUBLISHED in 1952 by epidemiologists Thomas McKeown and Reginald Record on fetal growth in multiple pregnancies reported that pregnancy length declines in a regular fashion as the number of babies increases. Whereas average length measured from the onset of the last menstruation is forty weeks for single births, it decreases to around thirty-seven weeks with twins, thirty-five weeks with triplets, and thirty-four weeks with quadruplets. Thus quadruplets are born six weeks earlier than singletons. In other words, with multiple pregnancies births are more likely to occur before the medically recognized threshold of thirty-seven weeks for premature babies. With singletons, fewer than one in ten is premature, but this number rises to five out of ten with twins and nine out of ten with triplets.

As is to be expected, birth weight decreases in tandem with pregnancy length. McKeown and Record recorded an average birth weight of 7.5 pounds for singletons, 5.25 pounds for twins, 4 pounds for triplets, and a little over 3 pounds for quadruplets. So individual quadruplets have less than half the birth weight of a singleton. Nevertheless, it is notable that the combined weight of the babies increases with the number born, from 7.5 pounds for a singleton through 10.5 pounds for twins, 12 pounds for triplets, and 12.5 pounds for quadruplets.

Interestingly, the weight of an individual fetus is seemingly unaffected by multiple pregnancies until about the twenty-seventh week. Beyond that point, the rate of growth decreases significantly as the number of fetuses increases. Retardation of fetal growth with multiple pregnancies is at least partially attributable to crowding in the womb, probably because of limits set by the placentas. Distension, or ballooning, of the womb is doubtless responsible for earlier birth with multiple pregnancies.

Multiple pregnancies with five or more fetuses—quintuplets, sextuplets, septuplets, octuplets, and nonuplets—are very rare, so authoritative figures for pregnancy lengths and birth weights are lacking. However, because both pregnancy duration and birth weight are tightly linked to the number of fetuses for the more common multiples, we can predict likely values. For octuplets, for instance, we can predict an average pregnancy length around thirty-one weeks and an average birth weight just above two pounds, with a combined weight of sixteen pounds. The first recorded octuplets in the United States were born in Texas in 1998. Six girls and two boys were born, but one infant later died. The remaining seven survived to celebrate their tenth birthday. Notorious "octomom" Nadya Suleman gave birth to the second known set of octuplets in the United States in 2009. The average birth weight of Suleman's eight babies, which all survived the first two years of life, was two and a half pounds. They were born after a pregnancy lasting a little over thirty weeks, almost ten weeks less than for a singleton. Both the average birth weight and the pregnancy length fit the predicted values quite closely.

Multiple births have become increasingly common in many countries since the 1970s. The main cause is mounting use of modern techniques of assisted reproduction, notably in vitro fertilization followed by embryo transfer. Hormone treatment is often used to increase the harvest of eggs for fertilization, and two or more embryos are commonly transferred to boost the chances of pregnancy. Reported cases of ten, eleven, and even fifteen fetuses all resulted from combined use of fertility medication and multiple embryo transfer. Among the nineteen reported cases of octuplets, at least thirteen were conceived with the aid of fertility drugs. Such trends are alarming because, as already noted, multiple-birth pregnancies result more often in premature babies. For this reason, fertility clinics are sensibly introducing tighter restrictions on the number of embryos transferred. Yet court cases revealed that the physician accountable for the octuplets born to Nadya Suleman, Dr. Michael Kamrava, actually implanted twelve embryos. Because his conduct was deemed irresponsible, Dr. Kamrava was ejected from the American Society for Reproductive Medicine, and the Medical Board of California revoked his license, although a judge later recommended reducing this sanction to five years' probation.

WHILE THE HUGE RANGE of variation in apparent lengths of human pregnancies, even with normal births of single infants, mainly reflects uncertainty about when ovulation occurs during the cycle, there can also be uncertainty about the actual cycle of conception. In the standard "egg timer" model, a woman has a series of regular monthlong menstrual cycles, each with midcycle ovulation, until she conceives. Once conception occurs, or so the thinking goes, menstruation automatically stops. Accordingly, the last menstruation can be taken as a reliable pointer to the onset of pregnancy several days later.

Things are not always that simple. Bleeding resembling menstruation may continue for up to three months after conception. There is tantalizing evidence that some kind of monthly cycle commonly persists into early pregnancy in nonhuman primates and even in other mammals. Regrettably, this is a little-studied topic. In one investigation of a pregnant gorilla that I conducted with colleagues at the Zoological Society of London, we identified the time of conception by monitoring hormones in daily urine samples. The hormonal profile definitively revealed that cyclical variation, with mating followed by bleeding a couple of weeks later, continued for at least two months into pregnancy. If bleeding resembling menstruation similarly continues into early pregnancy in humans, reliance on menstruation alone will yield a predicted birth date approximately one month too late.

Persistence of cyclical menstrual bleeding into early pregnancy may explain an observation that aroused a flurry of interest nearly a century ago. Occasional reports by gynecologists indicated that limited bleeding may occur close to the time of implantation in women, about a month after the last menstruation. This was seen as a possible sign of pregnancy. In the 1930s, reproductive biologist Carl Hartman noted similar bleeding around the time of implantation in his breeding colony of rhesus monkeys, and so bleeding close to implantation in both women and nonhuman primates became known as "Hartman's sign." A nice story, but likely not the right one. Hartman and others simply may have registered the relatively common occurrence of weak bleeding one month after the last menstruation; it is mere happenstance that this happens to be two weeks after conception.

Taking menstruation to indicate presence or absence of normal cycling is misleading in another way. A woman may falsely think that she is pregnant because she does not menstruate at the expected time. However, such irregularities are common. Even women who are not sexually active and hence cannot conceive occasionally have otherwise normal cycles without overt menstruation. In sexually active women, conception can occur without leading to implantation. Any pregnancy loss during the first eighteen weeks after conception is called a miscarriage. But all developmental stages between fertilization and implantation are tiny, so early loss easily passes unnoticed, which is why we know so little about it.

Careful monitoring, including hormone assays, is needed to distinguish early pregnancy loss from irregular menstruation. Miscarriage between four and eighteen weeks after conception is more noticeable than loss during the first month, which makes sense, as logic suggests that any natural mechanism for selecting against abnormal embryos should kick in early. In a landmark study published in 1980, gynecologist J. F. Miller closely monitored pregnancy loss in women. Conception was marked by detecting a pregnancy-specific hormone, human chorionic gonadotropin (hCG), in urine samples. Production of hCG begins at implantation, about ten days after fertilization. Miller and his colleagues recorded 152 conceptions, of which eighty-seven continued beyond the twentieth week, with all but two ending in live birth. Sixty-five detected pregnancies failed before midpregnancy, a loss rate of 43 percent. Fifteen of those failed pregnancies ended in clinically recognized miscarriages, but in fifty cases the sole evidence for pregnancy was detection of hCG in the urine. Because hCG is detectable only after implantation has occurred, the study completely missed any losses during the ten days between fertilization and implantation.

A year later, obstetrician Tim Chard presented a composite picture of losses after conception, combining various lines of evidence to conclude that on average, seven out of ten conceptions fail by midpregnancy. His estimations indicated a loss of 30 percent in the ten days between conception and implantation, when hCG first becomes detectable. If such early loss is accompanied by bleeding, it can easily be misinterpreted as menstruation. Chard identified a further loss of 30 percent between implantation and the end of the first month of pregnancy, after which miscarriage becomes clinically recognizable. In sum, clinically identified miscarriages account for only one in seven failed conceptions.

Substantial loss, exceeding two-thirds, between conception and the middle of pregnancy has been interpreted as a natural mechanism to block maturation of developmentally defective embryos. Aberrant chromosomes are a prominent feature of miscarriage. At least half of early pregnancy failures (those occurring during the first three months) can be attributed to chromosomal abnormalities, and the proportion may be even higher for losses between fertilization and implantation. In 1990, clinical geneticist Bernd Eiben published findings from placenta samples taken following 750 miscarriages, showing various chromosomal abnormalities in the cells of the developing placenta. Clearly, some kind of selection in early pregnancy occurs to limit chromosomal aberrations.

GOING BACK TO THE ISSUE OF TIMING, the scientific know-how needed to pinpoint ovulation and conception in women and in other primates has been available for decades. With suitable hormone assays, the rise in estrogens before ovulation, the midcycle spike of luteinizing hormone directly associated with ovulation, and the rise in progesterone after ovulation can all be used to monitor ovulation time. Conception is followed by a steep, persistent rise in estrogens and progesterone. While such complex and costly assays are rarely used to monitor normal pregnancies (they are largely reserved for problem cases or research), simple one-off hormone-based tests can easily be used to detect both ovulation and early pregnancy.

Throughout most of pregnancy, humans produce the hormone chorionic gonadotropin. Monkeys and apes also produce chorionic gonadotropin during pregnancy, so this is yet another feature that was probably present in the common ancestor of higher primates. As already mentioned, the hormone is first produced by the implanting blastocyst, starting about ten days after fertilization in humans. Later on, the placenta manufactures the hormone continuously until birth occurs. Standard pregnancy test kits are usually based on antibodies that detect hCG, which is eliminated largely intact in the urine.

Because production of hCG begins only ten days after conception, regular testing of urine samples with standard kits permits not only pregnancy diagnosis but also reliable identification of the cycle of conception. Anybody who wants to predict a birth date with greater confidence can easily do so by regularly using a standard pregnancy test kit. Secure identification of the cycle of conception reduces the range of uncertainty. That is a marked improvement over dates based on what seemed to be the last menstrual period.

Before we delve into details of embryonic and fetal development, a unique feature of early human pregnancy demands attention: so-called morning sickness. Mild to pronounced nausea, occasionally severe enough to provoke actual vomiting, affects around two-thirds of women in the first weeks after conception. Food aversions during pregnancy affect a similar proportion of women. In fact, nausea is frequently the first sign that a woman has conceived; because menstrual-like bleeding can continue into early pregnancy, morning sickness may strike before conception is suspected. It usually begins around the second week of pregnancy and generally stops by the twelfth week, although in rare cases it persists up until birth. The common name is misleading, as nausea is not especially prevalent in the morning—it could equally well be called "afternoon sickness" or "tedious all-day sickness." In an extreme form known as hyperemesis gravidarum, vomiting can be severe and lead to dehydration, weight loss, blood acidity, and potassium deficiency. Such debilitating effects occur in under 1 percent of pregnancies and require medical intervention.

Another, probably associated feature that mainly occurs in early pregnancy is a craving for unusual food or drink. Traditionally, nausea and food aversions or cravings have often been explained as negative side effects of hormonal changes during early pregnancy. However, this explanation is overly simplistic. Levels of certain hormones, notably estrogens, are actually far higher late in human pregnancy, when nausea is no longer likely and cravings are less frequent. Moreover, hormone levels have not been shown to differ between women with symptoms and those without.

Before exploring the evolution of menstruation, author Margie Profet focused on morning sickness. She suggested that the developing embryo is likely to be rather susceptible to toxins, and that an increased tendency to vomit might be an adaptation for mothers to eliminate any ingested toxins that could threaten embryonic development. A similar argument might be applied to food cravings—they could result from positive selection favoring the intake of nutrients needed by the embryo. Profet looked particularly at potential toxins in plant-based foods, singling out strong-tasting vegetables, alcohol, and beverages containing caffeine. But there is a much wider range of possibilities, including animal products, parasites, and disease agents. Unsurprisingly, Profet's "veggie hypothesis" received some harsh criticism. Epidemiologist Judith Brown and colleagues studied relationships between morning sickness, pregnancy outcome, and intake of supposedly harmful vegetables in an analysis of data for more than five hundred women. They found no statistical association between suspect vegetables and nausea or vomiting in early pregnancy. Moreover, intake of such vegetables was not correlated with adverse outcomes of pregnancy.

Another possibility is that morning sickness serves to protect both mother and embryo. In a comprehensive review, neurobiologists Samuel Flaxman and Paul Sherman examined various lines of evidence that support this interpretation. In the first place, symptoms peak when embryonic and early fetal development is most vulnerable to disruption by chemical agents, during weeks four to sixteen after conception. In addition, nine studies showed that women afflicted by morning sickness are significantly less likely to have a miscarriage than women who do not. Moreover, actual vomiting is associated with fewer miscarriages than nausea alone. By contrast, nausea and vomiting are not associated with stillbirths, which are losses later in pregnancy. The take-home message regarding ordinary morning sickness in early pregnancy seems to be: "Don't fight it, it's good for the baby."

Flaxman and Sherman showed that that morning sickness is often triggered by eating certain kinds of foods. Many pregnant women have aversions to alcohol, certain beverages (often those containing caffeine), and strong-tasting vegetables, especially during the first third of pregnancy. Yet the greatest aversions in pregnancy are to animal products—meats, fish, poultry, and eggs. Cross-cultural analysis by evolutionary psychologists Gillian Pepper and Craig Roberts revealed twenty traditional societies in which morning sickness had been recorded and seven in which it had never been observed. In the societies that lacked morning sickness, animal products were not major dietary items, and plants, mainly corn, were significantly more likely to be staples. Finally, Flaxman and Sherman also ruled out another proposal, that morning sickness reduces the frequency of copulation, thus preventing cramping of the womb that might lead to miscarriage.

Overall, it seems that vomiting may protect the embryo against disease agents and disruptive chemical agents, not just against defensive toxins produced by plants. Related to this, avoidance of parasites and foodborne infectious agents is very important for pregnant women. Their immune systems are weakened to reduce the likelihood that their bodies will reject the embryo. As a result, pregnant women are more susceptible to serious, often life-threatening infections.

What about our primate cousins? There is little evidence either for nausea or for unusual dietary cravings during early pregnancy in nonhuman primates. Any supposed negative side effects of hormonal changes during early pregnancy are hardly likely to apply only to our species. General arguments invoking positive or negative effects of dietary items on embryonic development should also be equally valid for other primates. On the other hand, it is entirely possible that nausea and/or cravings arose specifically during human evolution because our diet became so flexible, often including meat, thereby incurring greater risks. However, it is exceedingly difficult to propose a convincing evolutionary explanation with a sample size of one.

WE TURN NOW TO DEVELOPMENT of the offspring within the womb. As their name indicates, placental mammals all have a placenta, a special interface between the developing offspring and the surrounding inner wall of the womb.

The idea of a great chain of being (*scala naturae*) has a long history in philosophy. Since time immemorial, objects and beings have been arranged on an ascending staircase indicating progress. It goes without saying that humans always occupied the highest position on any staircase. In fact, some philosophers identified a series of intermediate stages such as angels and archangels to bridge a yawning gap in the great chain of being between humans and God. This ancient notion of an ascending staircase is so deeply embedded in Western thought that it is hardly surprising that it has influenced writings on evolution. One outcome of this insidious influence was that early scientific investigators often arranged mammals on an ascending scale. Egg-laying monotremes of course occupied the bottom rung. Marsupials were regarded as intermediate because they have live births but supposedly lack a proper placenta—instead, a shell membrane, seen as a carryover from an egg-laying ancestor, surrounds the developing offspring for most of pregnancy. All marsupials have brief pregnancies and give birth to tiny offspring whose development typically occurs in a pouch on the mother's belly. Placentals occupied the top rung because they have full-blown placentation and live births of relatively large offspring. Remember, though, that marsupials and placentals have evolved independently for at least 125 million years. Is it reasonable to take any modern marsupial as a frozen ancestor representing the starting point for both marsupials and placentals? In fact, several lines of evidence indicate that pregnancy length was actually reduced during marsupial evolution. Development in a pouch was seemingly favored at the expense of development within the womb.

Regardless of the evolutionary pathway followed by marsupials, it is true that modern placental mammals uniformly have a well-developed placenta. They also share a basic set of associated membranes. Whatever else happens, up until the time of birth the embryonic/fetal system is completely enclosed by a surrounding membrane, the chorion. The chorion encloses three other membranes that play specific roles in development. The first of these, the amnion, envelops and protects the embryo/fetus. It serves as a fluid-filled cushion that shields the developing offspring against any physical impact from outside. Rupture of the amnion and release of its fluid—commonly known as breaking water—accompanies birth. The remaining two membrane systems, the yolk sac and the allantois, supply nutrients to the embryo/fetus and remove waste products. To be more precise, it is the blood vessels of those two membrane systems that receive nutrients supplied by the mother and transfer waste products to her circulatory system for disposal.

One common feature of evolution through natural selection is the conversion of existing structures to serve new functions. This occurs at all levels, from anatomical features down to individual molecules. Tinkering is a major hallmark of evolution, and the yolk sac and the allantois provide superb examples. The common ancestor of land-living animals with backbones laid eggs containing, within a protective shell, everything needed for the offspring to develop. Many modern descendants from that ancestor—reptiles, birds, and monotremes—still lay eggs of that kind. The egg-laying mother provides nutrients for her offspring in the form of yolk, stored within its sac. Blood vessels splayed across the surface of the yolk sac gradually absorb those nutrients and ferry them to the developing offspring. Because an egg is almost a closed system once laid, apart from diffusion of respiratory gases through the permeable shell, the initial nutrient supply provided by the mother must suffice for the offspring's entire development. Moreover, waste products such as urea that cannot diffuse through the shell must be stored out of harm's way until hatching. Serving as a biological trash container was the original function of the allantois. Blood vessels lining the surface of the allantois gradually dump inconvenient by-products of development. At hatching, the waste-filled allantois is then discarded.

The egg of a marsupial or placental, by contrast, is adapted to develop within the womb using resources directly delivered by the mother. A substantial nutrient store is not needed, so there is very little yolk. In addition, because the mother's blood vessels carry away any waste products, the developing offspring does not need a built-in trash container. As vivipary evolved, the yolk sac and the allantois lost their original functions, but their blood vessels were recruited to serve a new function. Instead of absorbing a stored food supply, the yolk sac's vessels became adapted to exchange nutrients and waste products with maternal blood vessels in the wall of the womb. The blood vessels of the allantois, instead of dumping waste products, also became adapted to work alongside the yolk sac's vessels.

THOUGH ALL PLACENTAL MAMMALS have a well-developed placenta, there is considerable variation in detail, especially between orders of mammals. Some placenta types have a localized, usually disc-shaped interface, while others have a diffuse contact surface extending over most of the chorion. A localized placenta always invades the internal wall of the womb to some extent, although the degree of penetration varies. By contrast, a diffuse placenta is essentially superficial and noninvasive.

German physician Otto Grosser proposed in 1909 an influential three-way classification of placenta types: noninvasive, moderately invasive, and highly invasive. In a noninvasive, diffuse placenta, there is extensive contact between the chorion and the inner lining of the womb, but there is no significant breakdown of maternal tissue. With a moderately invasive localized placenta, some penetration of the inner wall of the womb occurs and maternal blood vessels come to lie directly against the chorion. In a highly invasive localized placenta, further penetration of the womb's inner wall breaks down the walls of maternal vessels and pools of blood directly bathe the chorion.

One of the great strengths of Grosser's classification system is that most mammal orders are characterized by only one type. For example, hoofed mammals, dolphins, whales, and pangolins all have a noninvasive placenta. Carnivores, elephants, sea cows, and treeshrews typically have a moderately invasive placenta. Rabbits and hares, most rodents, hyraxes, and elephant shrews all have a highly invasive placenta.

However, the order Primates stands out like a sore thumb. Uniquely, the two extreme kinds of placenta both occur within this single order. All lemurs and lorises have a noninvasive placenta, contrasting sharply with the highly invasive placenta universally found in tarsiers and higher primates. In fact, it was the highly invasive placenta of tarsiers that first led biologist Anton Hubrecht—the founder of comparative embryology—to propose in 1898 that they were linked to monkeys, apes, and humans. Features of placentation still count among the most convincing indicators that tarsiers and higher primates are descended from a common ancestor.

But how did the placenta evolve? To reconstruct its evolution in primates, we must first identify the starting point, the primitive condition that was present in the common ancestor of all placental mammals. We can then identify the likely type of placenta in the common ancestor that gave rise to all primates. At once a major handicap confronts us: Fossil evidence is unlikely to throw much light on the development of soft tissues, so we must rely on logical inference.

Many reproductive biologists have for some time believed that the problem of the evolving placenta had been satisfactorily solved, accepting the noninvasive condition as primitive. The chain of argument is fairly straightforward: Mammals evolved from egg-laying ancestors. The next step is the retention of the egg within the mother's body. At first the offspring inside the egg still depends almost entirely on its own supply of yolk, but the womb increasingly contributes directly to the offspring's development, initially with moisture and later in evolution with actual nutrients. For this to happen, the shell of the egg becomes thinner and increasingly permeable. Eventually diffusion between maternal blood vessels in the wall of the womb and blood vessels of the developing offspring will allow enhanced delivery of nutrients and removal of waste products. As the maternal supply of nutrients increases, the need for yolk in the egg declines. Similarly, as removal of waste products improves, the need to store them decreases. At some point this process leads to production of an egg with very little yolk that, after fertilization, develops within the womb using nutrients directly supplied by the mother. At the beginning, however, contact between the developing egg and the inner lining of the womb is necessarily noninvasive. Modern marsupials have been taken as a model for this stage of evolution. It seems only logical to infer that the placenta was still noninvasive in the common ancestor of placental mammals.

This stepwise view of evolution is bolstered by notions of efficiency. It is generally accepted that, as intervening barriers are reduced, resources are transferred with increasing efficiency from mother to offspring. Monotremes lay eggs, so their transfer efficiency is very low. Marsupials retain the developing egg within the womb, but a shell membrane surrounds it for most of pregnancy and there is usually no real placenta. So marsupials are seen as less efficient than placental mammals, which have a proper placenta. Among placental mammals, some have a noninvasive placenta with several barriers separating maternal blood vessels from those of the developing offspring. It is widely believed that the efficiency of transfer of maternal resources increases along with invasiveness of the placenta. According to this view, the most advanced condition must be the highly invasive placenta, which is also seen as the most efficient.

This is a classic case of the warping effects of the great chain of being. The outcome of this thinking has been wide acceptance of the notion that the placenta was still noninvasive in the common ancestor of placental mammals, as a carryover from an earlier marsupial-like condition. Consequently, evolution of placental mammals has often been reconstructed starting from the supposedly primitive and inefficient noninvasive placenta. Reproductive biologist Patrick Luckett has been a forceful advocate of this view, arguing that lemurs and lorises with their noninvasive placentation are primitive in every respect, while tarsiers and higher primates with their highly invasive placentation are advanced. Moreover, Luckett identified a progression of stages among the latter. In his reconstruction, tarsiers have the most primitive form of highly invasive placentation, New World monkeys are somewhat more advanced, Old World monkeys are even more advanced, and the most advanced condition of all is found in apes and humans. Comforting as it may be to have humans on the highest rung of the evolutionary ladder, this reconstruction unfortunately does not stand up to close examination once we consider pregnancy length and the state of offspring at birth.

Pregnancy length, as may be expected, generally increases with size. The longest pregnancy on record is that of the African elephant, averaging twenty-two months. We humans naturally think of our nine months as a pretty long haul, but we are medium-sized mammals and our long pregnancy is not unusual. So body size must be taken into account. To make meaningful comparisons between our pregnancy length and that of other animals, we must allow for a key difference in the condition of offspring at birth.

Anyone who has bred hamsters, hedgehogs, or mice knows that the mothers give birth to litters of poorly developed offspring. Newborn infants are pink, hairless little grubs at birth and their eyes and ears are sealed with membranes. By contrast, horses, cows, and chimpanzees give birth to single, well-developed offspring. They are typically born with a coat of hair, and their eyes and ears are open at birth. Largely thanks to zoologist Adolf Portmann, the crucial distinction between poorly developed, altricial offspring and well-developed, precocial newborns is now widely accepted. As a rule, altricial infants are born in a nursery nest in which they develop until they can move around independently. The eyes and ears open in the course of the nest-living phase. By contrast, most precocial infants, which are usually singletons, are able to move around independently from birth onward and have little need for a nest.

Portmann noted that pregnancies are relatively short in mothers of altricial offspring but long in mothers of precocial infants. An extreme example of the altricial condition is the common tenrec, a hedgehog-like mammal from Madagascar, which gives birth to a litter comprising some two dozen infants after a pregnancy lasting under two months. An example of the precocial condition at the other extreme is a three-ton female Indian elephant, producing a single precocial infant after a pregnancy lasting twenty-one months.

At any given size, a mother giving birth to a precocial infant has a pregnancy three to four times longer than a mother producing altricial infants.

For instance, a female cheetah—with a body weight quite close to that of an average woman—typically gives birth to four altricial cubs after a pregnancy lasting a little over three months. By contrast, a woman typically gives birth to a single well-developed baby after a pregnancy lasting three times longer. Yet the duration of human pregnancy is almost exactly what would be expected from our body size in comparison to other primates, our closest zoological relatives.

Interestingly, the relationship between pregnancy length and body size shows a fairly clear two-way split between altricial and precocial mammals, with almost no overlap. In most cases, placental mammals give birth either to several infants after a short gestation or to one infant following a long pregnancy. For some reason, the compromise solution of producing a medium-sized litter of infants after a medium-length pregnancy rarely occurs. Of course, there must be some kind of trade-off between offspring number and pregnancy length. As we have seen with multiple births in humans, the womb has a limited capacity, so a mother can produce either one quite large baby or several smaller ones. Yet the inevitable trade-off between number and size of newborns does not explain why natural selection has divided placental mammals into two clear categories with remarkably little overlap. This remains a prominent unanswered question in the evolution of mammal reproduction.

Portmann made another crucial observation: In each order of mammals mothers usually give birth to only one kind of offspring. For instance, marsupials, carnivores, insectivores, rabbits, and treeshrews typically have short gestation periods and produce hairless or sparsely haired altricial infants, while all hoofed mammals, dolphins, whales, elephants, bats, and primates typically have long pregnancies and produce hairy precocial babies. Rodents are unusual because some produce altricial offspring and others have precocial infants, although each suborder typically has a single newborn type. Because each major group of mammals has a typical newborn type, Portmann suggested that this must have been fixed early in evolutionary history.

As it happens, we have a rare advantage in exploring the evolution of newborn type, for the primitive condition is clearly evident. Portmann and colleagues discovered that in precocial mammals the eyes and ears become sealed during pregnancy and then open again before birth. Halfway through human pregnancy, for instance, the eyes and ears of the fetus are clearly sealed, reopening about three months before birth. In precocial mammals, such as primates, a preexisting nest phase was seemingly absorbed into fetal development as longer pregnancies evolved.

Comparison with birds provides valuable supporting evidence here. Birds, like mammals, have either altricial or precocial offspring. Altricial hatchlings of birds resemble newborn altricial mammals in being largely naked and having sealed eyes and ears. However, in birds it is the precocial condition that is primitive and associated with only rudimentary parental care. This means that there was no ancestral nest-living stage. Sure enough, as a precocial bird develops within the egg its eyes and ears do not seal over and then reopen, as in the fetus of any precocial mammal.

It is highly likely that ancestral placental mammals, like marsupials, gave birth to altricial infants. Therefore the condition seen in precocial infants is secondary and more advanced, accompanied by longer pregnancies and widespread reduction to single births. Knowing this, we can track the evolution of newborn type on a DNA-based evolutionary tree for mammals. If we assume that evolutionary change from the ancestral altricial condition occurred without any reversals, the switch to the precocial condition would have occurred independently in at least ten separate lineages. One of those ten lineages led to the precocial primates. To put it another way, absorption of an ancestral nest phase into pregnancy occurred no less than ten times separately. That is a truly remarkable example of multiple convergence in evolution.

We can now return to Portmann's proposal that a typical newborn type was established early in the evolution of every major group of mammals. Because altricial infants are primitive, this means that the shift from altricial to precocial offspring occurred early in the evolution of at least ten major groups of placental mammals. When Portman made his proposal, he probably did not expect that fossil evidence would one day provide dramatic confirmation of this.

One order of mammals now uniformly characterized by single precocial offspring is the group containing odd-toed hoofed mammals. This order includes horses, whose evolution can be traced back through the fossil record to the early Eocene, 55 million years ago. The almost 50-million-year-old Eocene site of Messel in southern Germany has become famous for its exquisitely preserved fossils. These include more than sixty skeletons of the early horse *Eurohippus*, which was the size of a fox terrier. Closer examination of eight of these horses revealed remains of a single, well-developed fetus, showing that horses already gave birth to a single precocial infant 50 million years ago. Messel has also yielded more than a hundred fossil bat skeletons. One of them, a specimen of *Palaeochiropteryx*, includes the remains of two well-developed fetuses. From this single specimen we can deduce that bats produced small numbers of precocial infants by about 50 million years ago. These findings strengthen the conclusion that the common ancestor of modern primates, which all produce precocial infants, was likely already precocial.

NEWBORN PRIMATES ARE WELL developed and generally fit the standard definition of precocial mammals. At birth, primates have a hairy coat, and their eyes and ears are usually open. They also have relatively long pregnancies and typically give birth to single offspring. But there are also important differences between newborn primates and precocial offspring of other mammals. Most precocial infants can move around independently soon after birth. This is true, for example, of hoofed mammals, dolphins, whales, and elephants. Primate offspring, by contrast, are commonly carried around, clinging to the fur of the mother or another group member, for an extended period after birth. So primate offspring are precocial, but not as independent as newborns of many other precocial mammals. As will be seen in Chapter 7, this has important implications for infant care.

Although newborn babies of nonhuman primates generally fit the precocial pattern, the human infant is often said to be altricial. Newborn human infants are indeed underdeveloped in comparison to those of other primates. Even so, describing them as altricial is misleading. First of all, we have a relatively long pregnancy, like other primates and in common with precocial mammals generally. And a human infant's eyes and ears are open at birth.

It might be thought that a newborn human baby does not fit the standard pattern for precocial mammals because there is generally very little hair. It turns out that loss of hair is clearly a secondary development in human evolution, highlighted by the title of Desmond Morris's book *The Naked Ape*. In fact, between the fifth and seventh months of pregnancy the human fetus is covered in fine, silky, pale lanugo, which is normally shed before birth. And if birth takes place eight weeks or more before the due date, lanugo, named after the Latin for "down," is usually still present. Strangely, those fetal hairs redevelop in severe cases of anorexia nervosa, possibly to reduce heat loss from an emaciated body. There is also a dominant genetic condition in which lanugo is retained into adulthood. As this trait runs in families, it has sometimes led to stories about "ape-men" or relict populations of Neanderthals clinging on in remote regions of Asia.

In most respects, then, while humans fit the standard pattern for precocial mammals, newborn human babies differ from infants of other primates in certain unusual features. Human infants are, relatively speaking, helpless: Our babies are far less mobile and far more dependent on parental care at birth than most primate infants. Unlike other primates, human newborns cannot grasp with their feet and are therefore unable to cling to the mother at birth. And dependency continues for a long time. Our unique ability to walk on two legs first emerges about a year after birth. Up to that point, the infant is at first immobile and thereafter crawls around on its hands and knees or scoots along on its bottom. We are the only primate species whose infants move around in an utterly different way from adults. All of these differences boil down to one main cause: The human brain is relatively less developed at birth. The next chapter will examine this difference in detail. Portmann was, of course, aware that human infants are unusual in comparison to other precocial offspring, and accordingly he described the human newborn as "secondarily altricial."

A CLOSE EXAMINATION of the topic of offspring size at birth leads to further insights regarding pregnancy length and offspring type. Because the womb has a limited capacity, it is only to be expected that altricial neonates will be relatively small, while precocial neonates will be comparatively big. Adjusting appropriately for mother's body size, analyses reveal that precocial offspring are indeed larger than altricial offspring at birth.

Within individual groups of mammals there are even finer differences between offspring. One finding, originally reported by biological anthropologist Walter Leutenegger in 1973, is that tarsiers and higher primates consistently have relatively larger newborns than lemurs and lorises. The difference is substantial. For a mother of any given body size, a newborn baby of a tarsier or higher primate is about three times bigger than one produced by a typical lemur or loris. It is tempting to conclude that this confirms the notion that an invasive placenta is more efficient. However, is this yet another example of the stork-and-baby trap, with confounding factors lurking in the background? An obvious test is to look at other mammals. If a noninvasive placenta is inefficient, all mammals with that kind of placenta should give birth to small offspring.

In fact, comparisons adjusted for body size reveal that newborn infants of primates—even those produced by tarsiers and higher primates—are generally smaller than those of other precocial mammals. The largest newborns, relative to mother's body size, are produced by hoofed mammals, dolphins, whales, elephants, and sea cows. Yet not one of these groups has a highly invasive placenta. Elephants and sea cows have a moderately invasive placenta, while hoofed mammals, dolphins, and whales all have a noninvasive placenta. Comparison across mammals reveals that newborn size is not connected with invasiveness of the placenta, clearly showing that a noninvasive placenta is not inefficient after all.

We can now return to our original question: How strong is the evidence that the noninvasive placenta is primitive for placental mammals? As shown above, numerous clues indicate that this oft-repeated conclusion is unjustified. Altricial offspring are most probably primitive among mammals, whereas precocial offspring are advanced. If the noninvasive type of placenta is primitive and inefficient, it should be associated with altricial offspring, but in fact the opposite is true. Mammals with a noninvasive placenta typically give birth to precocial infants. Altricial offspring are produced almost exclusively by mammals with a moderately or highly invasive placenta, although various mammals with an invasive placenta also produce precocial offspring. The only firm conclusion we can draw is that a noninvasive placenta is hardly ever associated with the primitive altricial type of offspring.

A complementary approach is to map the distribution of placenta types across a DNA-based evolutionary tree of placental mammals. A general guideline commonly applied in evolutionary studies is the parsimony principle: When we have to choose between alternative reconstructions, the version that requires the least overall change is most likely to be correct. We begin with a comparison of three alternative mammal trees, each starting with one of the three placenta types as the primitive condition. Next we count up the minimum number of changes required to produce the distribution of placenta types among modern mammals. Using this method, four independent studies, including my own, concluded that the ancestral condition for placental mammals was probably invasive. Nevertheless, conclusions diverge as to whether it was moderately or highly invasive. My own interpretation is that the placenta of ancestral placental mammals was moderately invasive. According to this view, evolution in primates would have taken two very different directions. In the common ancestor of lemurs and lorises, there was a shift to a less invasive placenta, resulting in the modern noninvasive condition. In the common ancestor of tarsiers and simians, by contrast, the placenta became more invasive.

The fact remains that as the placenta evolved, there must have been a noninvasive stage in which the developing egg was retained in the womb but had only superficial contact with its inner lining. That original noninvasive condition presumably existed long before the common ancestor of all modern placental mammals emerged. Remember that marsupials and placentals originally diverged at least 125 million years ago. By contrast, the common ancestor of modern placentals is more recent, perhaps dating back only 100 million years or so. That leaves a period of 25 million years or more for a moderately invasive placenta to evolve.

By this point the reader may well be thinking, "So what? What relevance does the evolutionary history of the mammalian placenta have for the modern human condition?" In my view, it is of utmost relevance. The longstanding inference that highly invasive placentation in humans is both outstandingly advanced and exceptionally efficient is wrong. Comparative evidence from placental mammals generally indicates that the noninvasive placenta can be just as efficient for converting maternal resources into offspring. If placenta type is not related to transfer efficiency of maternal resources, then what is its significance?

It seems highly likely that placenta type instead reflects a trade-off with immunological factors. As noted at the beginning of this chapter, an embryo or fetus developing within the mother's body possesses many foreign proteins coded by paternal genes. Therefore, special mechanisms must stop the mother's immune system from rejecting the developing offspring. But the more invasive the placenta, the greater the immunological challenge. So the real question is this: Why is any placenta invasive if maternal resources can be transferred across a noninvasive placenta with equal efficiency? An invasive placenta must have immunological advantages as well as disadvantages. Instead of pursuing the bankrupt notion that a highly invasive placenta must be more efficient, future research should focus more on the immune system.

If all goes well, human birth takes place at the end of a nine-month pregnancy. The next chapter will discuss some key aspects of the birth process itself. But two features that merit discussion here are the timing of birth and the fate of the placenta.

IN NONHUMAN PRIMATES, as was shown by biologist Alison Jolly in 1972, births most commonly occur when the mother is in the resting phase of her daily cycle. This pattern seems to be widespread among mammals generally, and it is undoubtedly advantageous for many tree-living primates. To avoid attracting predators, birth while exposed on the branches of trees must be discreet.

In contrast to many other day-active primates, human births—as any parent knows—are by no means confined to nighttime, but an underlying pattern does exist. Nowadays, of course, medical intervention in human births is so pervasive that it is difficult to obtain reliable information. However, early accounts documented birth timing before intensive manipulation became customary. Adolphe Quetelet, the first to report a human birth season, also reported a daily rhythm in the timing of birth, with a peak around midnight and a trough at midday. Some years later, pediatrician Edouard Jenny, in a 1933 review of more than 350,000 births in Switzerland over the period from 1926 to 1930, discovered that although births occurred at all times, they were most frequent between 2:00 and 5:00 a.m. and least common between 1:00 and 7:00 p.m. During the early morning peak, numbers of births were about 40 percent higher than during afternoon hours. Other studies support this pattern, including those by Americans Irwin Kaiser and Franz Halberg and Germans C. F. Danz and C. F. Fuchs. These findings are useful because they were conducted at a time when medical manipulation of human birth was limited. One particularly interesting finding is that duration of labor is significantly shorter during the hours when birth frequency peaks and longer at times when the incidence of births is lowest.

All available evidence indicates that if there is no intervention, a natural rhythm influences onset of labor and hence birth hour in humans. Spontaneous births occur around the clock but are more frequent during the early hours of the morning than in the afternoon and evening. In fact, closer examination reveals that it is the onset of labor, not the time of birth, that shows the clearest pattern of round-the-clock-variation. In all mothers, including first-timers, labor onset shows a prominent peak around 2:00 a.m. This pattern is perhaps no more than a carryover from ancestors that were adapted to give birth mainly at night as an adaptation to reduce predation. However, there is a strong possibility that the underlying rhythm remains biologically important and that early morning is the optimal time for human births.

Physiological processes underlying birth timing have been little studied, but gynecologist Maria Honnebier reported some interesting results in a 1994 paper based on her PhD thesis. She conducted a comparative study of timing of pregnancy and labor in women and in rhesus macaques. Various physiological features vary round the clock, such as body temperature, blood pressure, heart rate, levels of hormones associated with pregnancy, and mild muscle twitches (contractures) in the wall of the womb. In both rhesus macaques and women, contractures give way to full contractions as birth approaches. The hormone oxytocin plays a key part in this shift. Rhesus macaques show a clear twenty-four-hour rhythm in blood levels of oxytocin, and peak concentration coincides with the time when contractions occur.

Honnebier also carried out pulse tests with oxytocin to assess reactivity of the wall of the womb in rhesus macaques. The greatest response was found during nighttime hours. The same result was obtained with nine pregnant women tested between nineteen and thirty weeks of pregnancy. Thus there seems to be a physiological basis for the twenty-four-hour rhythm in frequency of spontaneous human births. As Jolly concluded in 1972, human birth hour is probably still subject to natural selection.

* * *

The fate of the placenta is also of particular interest with human births. Most female placental mammals eat the placenta soon after birth. As Benjamin Tycko and Argiris Efstratiadis aptly noted in a *Nature* commentary, they "have their cake and eat it too." The really striking thing about such "placentophagy" is that herbivores such as goats, which do not normally eat meat, eat the placenta just as readily as carnivores. Notable exceptions are seals and sea lions, dolphins and whales, and, for some reason, camels. Like most mammals, nonhuman primates all consume the placenta after birth regardless of normal dietary habits. In fact, with captive breeding colonies of primates, if a mother fails to eat the placenta, this is a fairly clear sign that she will not rear her infant. With great apes—chimpanzees, gorillas, and orangutans—mothers still fail to rear half of the infants born in captivity. The placenta is commonly left uneaten in such cases.

A long-standing explanation of consumption of the placenta is that it serves to avoid detection of a birth by predators. Alternative possibilities are that eating the placenta is in some way beneficial for the mother's health and/or that it stimulates the onset of maternal care. The placenta contains high levels of prostaglandins, which stimulate the return of the womb to its nonpregnant size and condition. It also contains traces of oxytocin, which reduces the stressful effects of birth and provokes contraction of muscles around the mammary cells, resulting in milk ejection.

In many human societies, the placenta has a special cultural significance, but it is only rarely eaten after birth. In a 1945 review based on the Cross-Cultural Survey at Yale University, American anthropologist Clellan Ford in fact reported that the placenta was buried in half of the human societies investigated. Nonetheless, in some societies it is believed that eating the placenta has beneficial effects such as alleviating postnatal blues and other complications of pregnancy. Consumption of the human placenta has been reported for Hawaii, Mexico, certain Pacific islands, and China. It is also an ingredient in some traditional Chinese medicines. But there is no hard scientific evidence to show that eating the placenta has any direct benefit for human mothers. Nevertheless, in the United States and Europe there has been a recent surge of interest in possible beneficial properties. To take one prominent example, American psychologist Jodi Selander founded the website PlacentaBenefits.info in 2006 to encourage mothers to consume the placenta after birth. She developed a proprietary method of placenta encapsulation based on processes used in traditional Chinese medicine. Her website states that consumption of placenta capsules speeds recovery after birth and reduces the risk of postnatal blues. However, if you aim to have your placenta encapsulated for later consumption, you may have an uphill battle. Reportedly, some hospitals will not readily hand over the placenta to a new mother eager to take it home and eat it. So plan your pregnancy carefully.

CHAPTER 5

Growing a Large Brain

In my youth, whenever I did well on a test my mother would tell me: "You get your brains from your father." This always seemed unlikely to me, as half of my genes came from her and the other half from my father. But later in life my research into brain evolution eventually revealed that she had modestly disowned the vital contribution to brain development that all women make. Mammals owe their brains to their mothers, at least with respect to investment of resources.

The brain is the body's command center, and one of our most vital organs. Brain evolution in mammals, in particular regarding the outstanding size of the human brain, has been much studied. Yet despite all the attention, the strong connection between reproduction and brain development has often passed unmentioned. The brain's high energy needs make it one of the most expensive organs in the body. In adult mammals, the energy needed to operate one ounce of brain tissue is about ten times as much as the average consumption of other living tissues. Although it accounts for only a fiftieth of body weight, the three-pound brain of an average adult human consumes around a fifth of the body's energy turnover.

For a developing brain, energy costs are even greater. Put simply, construction costs are added to basic running costs, increasing the heavy demand for resources. Like any operating system (other than regional governments), the body must achieve a balanced budget to be successful. Thus the higher resource costs for developing and operating a larger brain must be offset in some way. This adjustment can be achieved by some combination of using stored resources (drawing on reserves), increasing energy intake (augmenting income), or reducing costs for other parts of the body (downsizing in other departments). Balancing the budget commonly leads to some kind of trade-off between costs of the brain and other expenditures, notably those needed for reproduction.

AS LUCK WOULD HAVE IT, from the outset my research into primate evolution fostered an unusual combination of interests in reproductive biology and brain evolution. This mix of topics alerted me to cross-connections between these two areas. In particular, it drew my attention to the investment any mother makes in her developing offspring. As a general rule, evolutionary biologists tend to ask why any mammal species *needs* a large brain. Taking a different tack, I asked how any mammal is able to *afford* a large brain.

Biologically speaking, the crucial point is that in all mammals the mother provides most of the resources that her offspring needs for brain development. At first, during pregnancy, those resources are delivered across the placenta. Then, following birth, the mother nurses her offspring, providing further resources in her breast milk until weaning. In addition to being a conspicuous energy guzzler, the brain stands out in another respect compared to other organs: It grows rapidly early on and quickly reaches its target size. This course of development is understandable because the brain—the body's online computer—must be up and running once the offspring starts to move around independently. In mammals, most of brain growth, at least in size, has typically taken place by the time of weaning. Rapid early brain growth is followed by relative stagnation. This explains why the head of a newborn mammal is disproportionately large compared to the rest of the body and then becomes less prominent by adulthood.

In contrast to the brain, other major organ systems—heart, lungs, liver, kidney, digestive tract, muscles, and skeleton—grow fairly steadily and continuously from birth to maturity. In humans, for example, the brain has virtually reached its adult size by the age of seven, but the rest of the body continues to grow for another fourteen years or so. During later development, other organs increasingly outpace the brain. In a newborn human, the brain accounts for about a tenth of body weight, but by adulthood it is only a fiftieth. Because it has high energy demands and makes up 10 percent of body weight, a newborn human baby's brain actually consumes about 60 percent of total energy turnover. Such heavy consumption then gradually declines until it reaches the typical adult level of about 20 percent.

My comparisons among mammals revealed that adult brain size is significantly associated not only with energy consumption but also with pregnancy length. This discovery confirmed my gut feeling that brain development is closely tied to reproduction. The first clear clue to such a connection came from biologists George Sacher and Everett Staffeldt in 1974. They analyzed data for a representative set of mammals and showed that pregnancy length is more consistently associated with the brain size of a newborn infant than with its body size. Indeed, Sacher and Staffeldt concluded from this finding that the brain might serve as a pacemaker for the developing fetus. This proposal still awaits proper exploration, but brain size at birth is evidently tightly linked to pregnancy length.

The brain continues to grow after birth, fueled by resources delivered in the mother's milk. This means that suckling duration is also connected with the ultimate size of the brain. Yet the completed size of the adult brain is still significantly related to pregnancy length alone. Across mammals generally, the brain typically reaches about nine-tenths of its adult size by the time suckling stops. Therefore there can be no doubt that the mother is the provider of most of the resources needed for brain development.

In addition to its connection with pregnancy length, brain size in mammals also shows a clear statistical association with energy turnover. Recognition of the links between brain size, gestation period, and energy costs ultimately spawned my "maternal energy hypothesis." This hypothesis proposes that a mother's energy turnover during pregnancy directly influences the size of her newborn infant's brain. Other things being equal, the longer the pregnancy or the higher a mother's energy turnover, the more resources she can transfer across the placenta to promote brain development in her fetus. Similarly, the mother's energy turnover continues to influence her contribution to her infant's brain growth after birth, while she suckles the offspring. As the combined duration of pregnancy and suckling increases, the amount of time available for a mother to deliver resources for infant brain growth increases as well. In a real sense, then, when the offspring eventually reaches adulthood its completed brain size is largely attributable to maternal resources.

Primates obey this general rule, but they also show a unique relationship between brain growth and the development of the rest of the body. Throughout fetal development, primates consistently differ from all other mammals. George Sacher—this time in 1982— again made a seminal contribution to our understanding of brain development, discovering an important principle now known as "Sacher's rule." At all stages of pregnancy, a primate fetus consistently has about twice as much brain tissue as a similar-sized fetus of any other mammal. In other words, brain development is specifically privileged in primates.

However, the information available for Sacher's study was severely limited, as brain and body size have seldom been measured during fetal development, either in primates or in other mammals. Fortunately, there is an indirect way to test Sacher's rule. Because the difference between primates and nonprimates applies to all stages of fetal development, it is still present at birth. In contrast to fetal stages, data for brain and body size in newborn mammals are relatively abundant. Sure enough, when I assembled a large data set, analysis revealed that, for any given newborn weight, a primate has about twice as much brain tissue as a nonprimate. Sacher's rule was resoundingly confirmed.

Human infants are like other primates in this respect. Average brain weight of a newborn human baby can be accurately predicted from average body weight at birth using a scaling formula calculated for primates. A typical human baby weighing seven and a half pounds at birth has a twelve-ounce brain, pretty much what is expected in comparison to other primates. By contrast, a nonprimate mammal weighing seven and a half pounds at birth will usually have a brain weighing only six ounces or so. Because the brain is privileged throughout fetal development in all primates, their newborns are literally given a head start in life. Moreover, because this feature is universal among living primates, it must have been present in their common ancestor. The initial impetus for human brain expansion, and its connection with reproduction, are deeply rooted in our evolutionary past.

To understand brain growth after birth, we need to take account of the brain's state of development in the newborn. The previous chapter already introduced the fundamental distinction between poorly developed (altricial) and well-developed (precocial) offspring. Altricial mammals—such as mice, hamsters, hedgehogs, treeshrews, rabbits, and cats—have relatively short pregnancies. As expected, altricial newborns are small relative to the mother's body size and have small brains. Accordingly, a greater proportion of brain growth must take place after birth, typically while the infant develops in a nest.

As a general rule, in altricial mammals brain size increases approximately fivefold between birth and adulthood. Rapid brain growth continues approximately up until the eyes and ears open and the body is covered in fur. This developmental state roughly corresponds to birth in precocial mammals. After the eyes and ears have opened in altricial mammals the rate of brain growth is greatly reduced. The slower pace then continues until the brain reaches its adult size.

By contrast, newborn precocial mammals—primates, hoofed mammals, dolphins, and elephants—are relatively big and already have quite large brains, so less brain growth occurs after birth. The general rule for precocial mammals is that the brain approximately doubles in size between birth and adulthood. This degree of growth is far less than the typical fivefold increase for altricial mammals. Moreover, in precocial mammals, including nonhuman primates, the switch from rapid to slower growth of the brain typically occurs around the time of birth.

In this respect, however, humans are unique among mammals. As already noted, the relationship between brain and body size in a human fetus or newborn fits the general primate pattern. But after birth the pattern of human brain growth differs sharply from that of any other primate and any other mammal. Strikingly, human brain size increases almost fourfold after birth, rather than merely doubling as in other primates. This marked increase in human brain size after birth is combined with another unique feature: The rate of brain growth does not slow down around the time of birth, as it does in nonhuman primates and in other precocial mammals. Nor does the rate of brain growth slow down a few weeks after birth, as it does in altricial mammals. Human brain growth does not switch to a slower pace until about a year after birth.

To put it another way, the human brain continues to show a rate of growth as fast as that of a fetus for a year after birth. Continuation of a fetal pattern of brain growth for so long after birth explains why a newborn human infant is particularly helpless compared to other newborn primates. Zoologist Adolf Portmann fittingly noted that the length of human pregnancy should really be reckoned as twenty-one months: nine months inside the womb followed by another twelve months outside it. Anthropologist Ashley Montagu put it slightly differently, writing that in humans a nine-month period of normal pregnancy in the womb ("uterogestation") is followed by another nine months of fetus-like development in the outside world ("exterogestation").

It is clear that in a human newborn the brain and allied anatomical structures are immature compared to nonhuman primates and to other precocial mammals in general. This fact has several medical implications. One is an increased incidence of conditions affecting the ear, nose, and throat. Many of these conditions improve or are completely resolved during the first year of life. A prime example is inflammation of the middle ear (otitis) in human infants, which in severe cases can lead to hearing loss. Otitis is common in human infants because the Eustachian tube, which allows air to flow between the back of the throat and the middle ear cavity, is still relatively immature at birth.

RAPID BRAIN GROWTH during the first year of life is connected with another unusual feature of newborn human infants: their striking plumpness. In an average human newborn weighing some seven and a half pounds, fat tissue accounts for over a pound, around 14 percent. Our babies are among the plumpest found among mammals. Human babies at birth look markedly different from the scrawny newborns of other primates, such as chimpanzees and rhesus monkeys. The proportion of fat tissue in a newborn human matches that in mammals living under arctic conditions and actually exceeds the level found in baby seals. As anthropologist Christopher Kuzawa has shown, a newborn human has about four times as much fat as expected for a standard newborn mammal of the same body size. In fact, the proportion of body fat in a human baby increases further over the first nine months after birth, building up to about a quarter of body weight. During that period, around 70 percent of the energy allocated to growth is used to deposit fat. In short, healthy babies do not lose their baby fat after birth but consolidate it and maintain it for up to three years. A mother's investment in building up her infant's fat reserves continues long after birth, largely thanks to nursing.

A standard explanation for our plump babies has been that natural selection favored an increase in body fat to offset the loss of insulating body hair. It is known that the optimal temperature for a human infant kept in an incubator is about 90°F, so cooling could be a problem. Baby fat is distinctively distributed, being mainly located just beneath the skin. In contrast to adult fat stores, there is relatively little fat in the belly cavity. Anthropologist Bogusław Pawłowski supported this view, arguing that various features of the human newborn evolved in early *Homo* to counter excessive cooling during nights spent sleeping in open savannah. Those features include relatively large size as well as a greater proportion of subcutaneous fat. A sleeping human infant is also unusual in being able to actively regulate its own body temperature.

However, Kuzawa's studies yielded only weak evidence for the role of subcutaneous fat proposed by Pawłowski. Kuzawa went on to explore a more likely explanation for our exceptionally plump babies: increased fat reserves as a crucial energy buffer. This would be particularly advantageous during the period of rapid brain growth in the first year of life. It could offset any disruption in the flow of resources to the growing infant. Going a step further, a 2003 paper by two nutritionists, Stephen Cunnane and Michael Crawford, argued that plump babies are the key to evolution of the large human brain, and not only because of energy supply. About half of the brain consists of fat, and a baby's fat reserves contain special fats—long-chain polyunsaturated fatty acids (LCPUFAs)—that are essential for normal brain development. Calculations indicate that LCPUFAs present in baby fat at birth should be enough to fuel three months of brain growth. Cunnane expanded on this theme in his 2005 book *Survival of the Fattest*, in which he described the normal human newborn as "positively obese." Deposition of fat in the human fetus takes place only during the last third of pregnancy; almost no fat is present during the first six months. As a result, fat reserves are well below normal in premature babies. A baby born five weeks early has only half the usual amount of fat, and a baby born ten weeks early has less than a sixth. With such early preemies the ribs and chest muscles stand out because there is so little fat tissue beneath the skin. Insufficient fat deposits mean that preemies are not well buffered for the rapid brain growth that takes place after birth. Although normal brain growth can nevertheless occur given adequate nutrition, it is vital to recognize the special needs of premature babies. Cunnane aptly describes stored fat in the newborn human as "insurance."

As previously discussed, aside from brain development, newborn human infants are in most other respects like other precocial offspring. Human babies are "altricial" only in a special sense: The brain is underdeveloped at birth, compared to its completed adult size, and continues to grow rapidly after birth. Portmann's description of the condition of the human newborn as "secondarily altricial" is fitting. It is the mismatch between brain development and other aspects of their condition at birth that makes our babies special. One reflection of this is that the bones of the skull are not fully developed at birth, leaving gaps known as fontanelles on the crown and on the sides. In human babies, these gaps, which have closed by birth in monkeys and are quite small in newborn apes, usually do not close until eighteen months to two years after birth.

Extension of a fetal pattern of rapid brain growth into the first year of life has some important implications. In the first place, the switch from fast to slow brain growth in the human infant is not associated with opening of the eyes and ears, as is otherwise typical for mammals. Human infants are unusual in having open eyes and ears during a yearlong period of fetal-type brain growth after birth. This extraordinary feature allows a human baby to interact with the environment with a relatively immature brain. As all parents know, human infants learn a great deal during the first year of life and already engage in sophisticated social exchanges. This feature was hugely influential in the gradual emergence of greater adaptability and behavioral flexibility during human evolution.

In this connection, it is no coincidence that the species-typical way in which human beings move around—walking upright—does not develop until about a year after birth. For a few months before striding develops, a human baby moves around in a distinct way, such as crawling on hands and knees or scooting. Here is another feature in which we stand out, since other primates move around in a species-typical fashion from the outset. It is also notable that active use of language does not begin until a human baby's second year of life, although infants busily learn many basic rules of communication during their first year. In sum, while human infants continue a fetal pattern of brain growth for a whole year after birth, they are able to interact actively with their surroundings.

But this uniqueness leads us to an important question: Why did this special human pattern of brain development evolve? The answer is quite straightforward. After nine months of development inside the womb, the brain size of the human fetus reaches the upper limit for safe passage through the pelvic birth canal. If the birth canal did not impose a size constraint, pregnancy might last about twenty-one months, continuing for a year or so beyond the usual time of birth to ensure full brain development.

It is far more efficient to develop brain tissue by transferring resources directly across a placenta. After birth, the mother must first convert her resources into milk, which is then transferred to the infant for digestion. This to some extent explains why human development in the womb is pushed right up to the limit allowed by the dimensions of the birth canal. Dimensions of newborn baby heads reflect this. As a general rule in biology, variation in any dimension within a species fits a typical bell curve or normal distribution. The central peak of the bell curve is the average value and smaller or larger values decrease progressively in a mirror-image fashion on either side of the average. The head circumference of a human newborn is a striking exception to this typical distribution. Head dimensions that are below the average decrease in the expected way, but above the average there is an abrupt decline in the largest head sizes. This reduction in upper-end variation is a sure sign of the filtering action of natural selection, eliminating head sizes that are too big for a safe birth.

Comparison with great apes, our closest zoological relatives, reinforces this conclusion. They all have shorter gestation periods: thirty-five weeks in orangutans, thirty-seven weeks in gorillas, and only thirty-three weeks in common chimpanzees. Female chimpanzees and orangutans are generally smaller than women, while female gorillas are somewhat bigger. Our true gestation period of thirty-eight weeks is between one and five weeks longer than in great apes. Body size does not explain this difference. And it certainly does not account for the dramatic difference in size at birth: Newborn orangutans and chimpanzees weigh approximately four pounds and baby gorillas about four and a half pounds, compared to an average of seven and a half pounds for newborn human infants. To put it another way, newborn body size is around 3 percent of mother's body mass in monkeys and apes and almost 6 percent in humans.

Thus a human baby weighs almost twice as much as a great ape infant at birth. Remember Sacher's rule: Brain and body size at birth fit a standard relationship across primates. As a human newborn is about twice as big as any great ape at birth, its brain size is correspondingly almost twice as big as well. This disparity tells us two important things. First, women must invest considerably more in fetal development than any great ape to produce our larger-bodied, larger-brained newborns. Second, it shows that humans really do push right up against the limits as regards brain and body size at birth. This explains why human birth is such a drawn-out, hazardous process.

Unusual challenges in human birth are also reflected by the difference between men and women in the shape of the pelvis. In monkeys and apes males and females do not obviously differ in pelvis shape. By contrast, a woman's pelvis is quite different from a man's. The width of the pelvis is about the same in the two sexes, but it is relatively wider in women because they have smaller bodies. Think of body shape. In women, the hips are generally wider than the shoulders, whereas the opposite is true for men. Internally, the lower end of the spine is shifted backward in women so that it does not bulge into the pelvic canal, which is smoothly rounded. In men, such a backward shift of the spine is unnecessary and the pelvic opening is heart-shaped. In addition, the joint between the left and right halves of the pelvis at the front, in the pubic region, is shorter in women than in men and the angle below is wider. There are numerous other differences, making it easy to tell whether a human pelvis is male or female. Biological anthropologists who investigate human skeletons from archaeological sites or crime scenes—skeleton detectives—will look at the pelvis first if they need to identify an individual's sex. Sex differences in the pelvis also result in different walking styles. When women walk, their hips sway because the pelvis is tilted more toward the front and moves up and down more obliquely. Movements at the hip and knee joints also differ.

The proposal that pelvic dimensions limit head size in the human newborn is quite logical and can be independently verified by an indirect test. All we need is to find a mammal that produces a large-bodied, large-brained newborn comparable to a human baby yet has no constraint on its birth canal because it has no bony pelvis. Thanks to the wonders of biological diversity, such mammals do exist. The ancestors of dolphins and whales secondarily returned to life in water. As a result, the entire hindlimb girdle—including the pelvis—became redundant and a few bony splints are all that is now left. And some dolphins happen to be reasonably close to humans in body weight, pregnancy length, and adult brain size. A dolphin with no pelvis can give birth to a particularly large baby with a brain more than twice as big as the twelve-ounce brain of a human newborn. But the brain of a dolphin merely doubles in size between birth and adulthood, rather than quadrupling as it does in humans. Water-living, pelvis-less dolphins have no need to extend fetal brain development into postnatal life.

Because the head of a newborn human is already so big compared to the size of the birth canal, its passage through the pelvis is fraught with difficulty. Birth is eased to some extent by specific action of the hormone relaxin. Production of relaxin, an insulin-like hormone produced by the ovaries, placenta, and breast, peaks late in pregnancy. Among other things, the hormone softens up the ligament that binds the left and right halves of the pelvis together at the front. It also relaxes the pelvic musculature, rendering the pelvic canal a little more flexible. The fontanelles between the main bones of the skull play a part in easing birth as well, allowing flexibility in the shape of the newborn's head. Nevertheless, passage of the human infant through the pelvic canal during birth is still a remarkably tight squeeze. Some kind of physical obstruction arises during about one in five human births.

Difficulties in the human birth process have been remarked upon through the ages. In the Bible, for instance, the following divine punishment for original sin was meted out to Eve and every woman thereafter (Genesis 3:16): "I will greatly multiply your pain in childbirth. In pain you will bring forth children." This biblical connection between eating forbidden fruit from the Tree of Knowledge and painful childbirth is intriguing in view of the clear link between large brain size and challenging human births.

To pass through the pelvic canal, the human newborn undergoes a complex pattern of rotation that is highly unusual among primates. The tortuous pathway followed is not only due to the newborn infant's relatively large head and broad shoulders. It also reflects changes in shape and orientation of the adult human pelvis for upright two-legged walking. These two factors combined in evolution to produce a veritable obstacle course for the human infant during birth.

Problems caused by a relatively large newborn head are not confined to humans. As anatomist Adolph Schultz originally noted, newborn head size can also create difficulties at birth for certain nonhuman primates. Once again, body size is a key factor that must be taken into account. As is to be expected, across primate species average newborn body size increases with average mother body size. But, as for many other biological features, scaling is not simply proportional. Across species, newborn size increases at a slower pace than mother's body size. In other words, as the mother's body size increases, the size of her newborn represents a smaller and smaller fraction of her weight. In comparison to mother's size, small-bodied monkeys have large infants, medium-sized monkeys and gibbons have moderate-sized infants, and great apes have small infants. For instance, squirrel monkeys—large-brained, small-bodied New World monkeys—have relatively large newborns that fit tightly in the birth canal. But the monkey pelvis, unlike that of humans, has not been radically modified for upright walking. So birth is relatively straightforward despite the snug fit.

In humans, birth of a large-headed newborn takes place through a pelvis reconfigured for two-legged striding. Moreover, the birth canal is tortuous because the inlet into a woman's pelvic canal is largest from side to side, while the outlet is largest from back to front. As a result, a two-stage turning sequence is needed for the newborn to pass safely through the pelvis. Anthropologist Karen Rosenberg described the special features of human birth in a 1992 paper, showing that as the human newborn's head enters the inlet of the pelvic canal, it is already rotated so that its long axis is oriented from side to side rather than from back to front, as is typical for nonhuman primates. Then, during its passage through the pelvis, the infant's head is rotated even more to fit the front-to-back orientation of the long axis of the pelvic outlet. Its face is usually pointing toward the mother's rear as it emerges. In other primates, the baby's face is typically directed forward as it passes through the pelvis, and there is no rotation at all.

The human newborn's large head is not the only thing that makes birth problematic. The infant's shoulders are also quite wide relative to the birth canal, so additional juggling is needed for them to pass through. This is why the head is turned to one side as the shoulders pass through the birth canal. The shoulders, too, make a tight fit, so jamming is an additional hazard during human birth, occurring in up to one in a hundred cases.

The distinctiveness of the human birth process has perhaps been somewhat exaggerated. In 2011, primatologist Satoshi Hirata reported that in a captive colony of chimpanzees close-up video recordings showed that the head was rotated to face backward in all three births that were monitored. The head and body of the newborn rotated to face forward after the head had emerged. As Hirata and his coauthors noted, the actual birthing process has been little studied in nonhuman primates, so rotation in the birth canal may occur in other species. However, if birth with the newborn's head facing backward occurred regularly in standard laboratory primates such as rhesus monkeys, squirrel monkeys, and common marmosets, surely it would have been noticed. Moreover, the uniqueness of human birth is not limited to rotation but also has to do with head size and adaptations for upright striding. For these reasons, two-stage rotation is obligatory.

All of these special features combine to make human birth a drawn-out and difficult process, and it is hardly surprising that medical professionals call it "labor." One multicultural study, published in 1999 by gynecologist Leah Albers, analyzed the duration of labor for more than 2,500 full-term births. All cases involved low-risk mothers who gave birth without medical intervention under the care of nurse-midwives in hospitals. Birth took an average of almost nine hours for first-time mothers and around six hours for mothers with previous births. In extreme cases, birth took as long as twenty hours.

By contrast, in nonhuman primates birth is relatively rapid and straightforward. Anthropologist Wenda Trevathan reviewed birth duration in primates in her 1987 book *Human Birth: An Evolutionary Perspective* and found that, as expected from the small size of newborns compared to the girth of the mother's pelvis, birth is usually relatively uncomplicated in great apes. Orangutans, chimpanzees, and gorillas all give birth in a couple of hours. The unexpected backward-facing births of chimpanzees reported by Hirata and colleagues do not entail prolonged birth duration, as in humans; to the contrary, they may well be the outcome of a loose fit within the birth canal. As in great apes, birth usually takes a couple of hours in monkeys as well. But with small-bodied mothers labor can be more difficult because of a tighter fit between the baby and the birth canal. Certain small-bodied, large-brained monkeys—notably squirrel monkeys—have been reported to have relatively difficult births, although they also last only a couple of hours. In any event, humans are the only primates that have such protracted, difficult births at such a large body size.

Wenda Trevathan and Karen Rosenberg also noted that complex human births, with their associated risks, make some kind of assistance almost obligatory. Birth of a backward-facing newborn and the danger of jamming present major challenges. In addition to providing general assistance during birth, a midwife can intervene to avoid problems that may arise. In about a third of human births the umbilical cord becomes wrapped around the baby's neck because of the complex pattern of rotation. In most cases this is not life-threatening, but occasionally the cord tightly constricts the baby's neck. If corrective action is not taken quickly, the infant can be strangled. Surveillance and prompt intervention by a midwife can ensure that the umbilical cord does not create a serious threat during birth.

As brain size gradually expanded over the course of human evolution, it must have posed an ever-increasing challenge for the birth process. Over the past 4 million years, the size of the brain approximately tripled, increasing from about a pound in *Australopithecus* to roughly three pounds in modern humans. Modern great apes and humans have quite similar pregnancy lengths, so we can reasonably assume that their common ancestor had a relatively long pregnancy as well, about eight months. As brain size increased in the hominid line, pregnancy gradually increased to reach the nine months typical for humans today.

In the earliest stages of hominid evolution, as in modern great apes, dimensions of the pelvic canal probably imposed little constraint on birth. When the switch to upright, two-legged striding in human evolution occurred, it changed both the size and the shape of the pelvic canal. As brains became bigger, the pelvis increasingly constrained the size of the newborn's head. At some point it became necessary to postpone part of fetal brain growth until life after birth. This was the only way to bridge an expanding gap between the largest newborn brain size permitted by pelvis dimensions and the completed size of the adult brain.

Australopithecus, the earliest well-documented hominid, existed between 4 million and 2 million years ago. At about a pound, average adult brain size was still relatively small. In modern nonhuman primates, the brain at birth is generally about half the size of the adult brain, and if *Australopithecus* still fitted this pattern, the newborn would have had an eight-ounce brain. It is unlikely that its size would have posed a major problem. However, various studies have indicated that the pelvis of *Australopithecus* had perhaps already begun to constrain the birth process because adaptations for upright two-legged walking were already under way. In particular, the pelvis had become relatively broad and low-slung, contrasting with the tall, narrow pelvis of great apes. Remodeling of the pelvis changed the shape of the bony birth canal in *Australopithecus*, making it wider from side to side and narrower from front to back. Perhaps this change in shape prevented the newborn's head from passing straight through the birth canal in the typical nonhuman primate fashion. Instead, some obligatory rotation of the baby's head may have been needed to align its long axis with the largest dimension of the entry into the pelvic canal, as in the first-stage rotation seen in humans. Some authors have even suggested that birth in *Australopithecus* was as complex as in modern humans, although that is unlikely. Additional rotation of the newborn's head during birth was probably unnecessary.

Unfortunately, the size of the brain and body in newborn *Australopithecus* is not documented by fossil evidence. Instead, various estimations have been made using comparisons with living species. To roughly assess brain size at birth, for example, we might take a graph for monkeys and apes showing the size of the newborn's brain in relation to the mother's body size. The same approach could be taken to estimate newborn body size. Yet we are faced with a tricky problem. As mentioned, compared to monkeys and apes modern human babies are far larger than expected, with appreciably larger brains. Thus human newborn brain and body size would be greatly underestimated if calculated from such a graph. This would not matter if *Australopithecus* did not differ from modern monkeys and apes. On the other hand, if *Australopithecus* had already started to evolve toward the modern human condition, its newborn brain and body sizes would be underestimated similarly. The whole point of the calculations is to find out whether *Australopithecus* was ape-like or human-like, so we are locked in a vicious circle.

In an attempt to break that circularity, anthropologist Jeremy DeSilva came up with an ingenious suggestion. Although human newborn brain and body sizes are larger than expected in comparison to monkeys and apes, the relationship between adult brain size and newborn brain size is more consistent across species. Therefore we can estimate newborn brain size for *Australopithecus* from adult brain size, which is well documented in the fossil record. Sacher's rule reflects a consistent relationship between brain size and body size in newborns, so once we have an estimate of newborn brain size we can use it to calculate newborn body size. Taking this approach, DeSilva concluded that *Australopithecus* had larger-brained, bigger-bodied babies than a modern ape of comparable adult size. This, in turn, indicates that birthing difficulties indeed might have begun to emerge even in *Australopithecus*.

Yet DeSilva's approach does not entirely avoid circularity, although it does reduce it. Between birth and maturity, human brain size almost quadruples, whereas brain size merely doubles in an average nonhuman primate. Because of this, if we try to calculate newborn brain size from adult brain size in humans, using a graph for monkeys and apes, there will be some overestimation. Nevertheless, the fact remains that *Australopithecus*, which is actually somewhat smaller than a chimpanzee in body size, ends up with estimates for neonatal brain and body size that differ from those of chimpanzees.

Although there is no fossil evidence to indicate brain size in newborns, fairly complete skulls of two three-year-old *Australopithecus afarensis* from Ethiopia and of a four-year-old *Australopithecus africanus* from South Africa have been discovered. The brain sizes of these immature australopithecines are similar to those of chimpanzees of the same age. But adult chimpanzees have smaller brains than adult *Australopithecus*, about thirteen ounces instead of a pound. This means that a smaller proportion of brain growth was achieved in three-year-old and four-year-old *Australopithecus* than in chimpanzees of the same age. So there is indirect evidence that more brain growth occurred after birth in *Australopithecus* than in chimpanzees, hinting at the beginnings of a pelvic constraint.

Advanced hominids belonging to the genus *Homo* first appear in the fossil record around 2 million years ago. Brain size steadily increased among successive *Homo* species, so challenges in the birth process would have increased in parallel. In *Homo habilis*, for example, average adult brain size had already increased by a third compared to the norm for *Australopithecus*, from a pound to about twenty-one ounces. If *Homo habilis* followed the typical pattern for nonhuman primates, newborn brain weight would have been about ten and a half ounces. This is only 15 percent less than the twelve-ounce newborn brain size of modern humans. Because *Homo habilis*, like *Australopithecus*, was relatively small-bodied, the larger head of the newborn was probably even more subject to limits imposed by the birth canal. Unfortunately, the skeleton of *Homo habilis* is poorly documented in the fossil record, so we must await further discoveries to assess the extent of difficulties at birth.

Like *Homo habilis*, *Homo erectus* dates back to around 2 million years ago, but the latter was closer to modern humans in body size. *Homo erectus* also had a bigger brain than *Homo habilis*, with adult brain size averaging about two pounds. If *Homo erectus* followed the general rule for nonhuman primates, brain size in the newborn would have been about a pound—four ounces more than the twelve-ounce brain of a modern human newborn. Pelvic dimensions in *Homo erectus* are quite close to those in modern humans, so this hominid species doubtless showed some extension of a fetal pattern of brain growth into postnatal life. Development in the womb in *Homo erectus* was most likely pushed to the limit, as in modern humans. Therefore this species probably also gave birth to newborn infants with a twelve-ounce brain. If pregnancy lasted nine months, as in modern humans, prolongation of a fetal pattern of brain growth into the first three or four months of life after birth would have been needed. Accordingly, we can speculate that newborn *Homo erectus* already showed a partial version of the "secondarily altricial" state of newborn humans today. Increased helplessness of infants during the first few months after birth would have made intensive parental care obligatory. At the same time, birth of infants with a brain that was still relatively immature would have begun to open up opportunities for increased behavioral flexibility and early social learning.

Limitations of the fossil record once again constrain our interpretations. Thus far, only a single fossil specimen indicates brain size in a young individual at this stage of evolution: a partial skull of a *Homo erectus* infant discovered in 1936 at the 1.8-million-year-old Mojokerto site in Java, Indonesia. Making matters worse, the specimen preserves only the braincase and lacks the face and teeth, making it difficult to estimate the infant's age at death. However, in 2004 anthropologist Hélène Coqueugniot and colleagues performed a detailed study of the Mojokerto braincase using CT scanning. Combining results from three different age indicators, they concluded that the infant was about a year old. Its brain weight came out at a little over twenty-three ounces, roughly three-quarters of the average brain size of slightly more recent adult *Homo erectus* from Java. Coqueugniot and her team concluded that brain growth was already well advanced in the Mojokerto infant and that there had been little evolution toward a "secondarily altricial" condition. Two years later, though, anthropologist Steven Leigh came to a different conclusion. He noted that, for a one-year-old, the brain size of the Mojokerto infant fell within the lower end of the modern human range rather than into the range of chimpanzees. Moreover, one of the age indicators used by Coqueugniot was recent closure of the fontanelle on the skull roof. In fact, closure of that fontanelle at an age of about a year itself indicates that the brain of the Mojokerto infant must have been relatively immature at birth. So the limited and uncertain evidence that is available does suggest that *Homo erectus* was moving toward the modern human condition.

NEANDERTHALS (*Homo neanderthalensis*) and *Homo sapiens* both descended from *Homo erectus*. The lineages leading to these two advanced sister species diverged at least half a million years ago and perhaps even earlier. Brain size increased from two pounds to three pounds in both Neanderthals and modern humans after they diverged from *Homo erectus*. Most of that one-pound increase in brain size occurred independently in the two lineages, as is reflected by marked differences in brain shape between *Homo neanderthalensis* and *Homo sapiens*. Any further extension of a fetal pattern of brain growth into life after birth, beyond the condition present in *Homo erectus*, would have taken place separately in the two species as well. This means that the need for more intensive parental care would have increased as a parallel development in Neanderthals and modern humans. By the same token, the scope for greater behavioral flexibility and early social learning in young infants must have increased independently in the two lineages.

Because closely related mammal species belonging to the same genus consistently have similar gestation periods, it is safe to assume that pregnancy in Neanderthals lasted nine months, just as in modern humans. Moreover, it now seems that the bony pelvic canal of female Neanderthals was quite similar to that of modern women. Fragments of an adult female

Neanderthal pelvis were discovered at the Tabun site in Israel more than eighty years ago, but its overall shape remained uncertain until computerized techniques became sophisticated enough to carry out virtual reconstructions. Using this approach, anthropologists Timothy Weaver and Jean-Jacques Hublin found that areas of the pelvic inlet and outlet are actually quite similar to those of *Homo sapiens*. Nevertheless, the shape of the birth canal is distinctly different. The inlet and, particularly, the outlet are wider in the Tabun pelvis as compared to the modern human one. Because of this, a second-stage rotation of the head for exit from the birth canal would have been unnecessary in the Neanderthal, and the baby would have emerged with its head facing sideways. A marked shift to the modern birth pattern, with the baby's head undergoing a second rotation to face backward at emergence, seemingly occurred after Neanderthals and humans diverged.

The fossil record for Neanderthals is considerably better than for earlier hominids, and two fairly complete skeletons of newborn individuals have been discovered. The first is from the Le Moustier cave site in Dordogne, France. In 2010, anthropologist Philipp Gunz and colleagues included the Le Moustier newborn in a detailed study of brain development after birth. The analyses showed that Neanderthals and modern humans are quite similar with respect to brain development at birth, but they follow quite different trajectories thereafter. Whereas the brain of Neanderthals takes on an elongated form as it develops, the brain of *Homo sapiens* becomes globular in shape. The second newborn Neanderthal specimen, found in 1999, was discovered in Mezmaiskaya Cave in Russia. In 2008, anthropologist Marcia Ponce de León and colleagues performed virtual reconstructions of the Mezmaiskaya newborn, concluding that brain size at birth was similar in Neanderthals and modern humans. In fact, in both the Mezmaiskaya and the Le Moustier specimens, the newborn's brain was similar in size, at about fourteen ounces. This weight is somewhat above the average for modern humans, so it seems that female Neanderthals did not have an easier time at birth.

AT BIRTH, all primates have larger brains for their body size than all other mammals. Yet by the time adulthood is reached, the distinction between primates and other mammals is fuzzier because of development after birth.

It is often claimed, even in academic texts, that adult primates have bigger brains than other adult mammals. This claim is misleading. It is, of course, untrue for absolute brain size. An adult elephant has a brain four times bigger than the average human brain, and a sperm whale's brain is, at sixteen pounds, the largest of any mammal. Reliable conclusions are possible only if we scale brain size to body size across mammals. Eugène Dubois, the discoverer of *Homo erectus*, was among the first to recognize this at the end of the nineteenth century. Since then, the relationship between brain size and body size has been much discussed. Two works that broke ground in this area were Harry Jerison's 1973 compendium *Evolution of the Brain and Intelligence* and John Allman's 1999 book *Evolving Brains*.

A rudimentary measure to compensate for size effects when comparing brain size between species is a simple ratio of brain size to body size. For instance, my brain is 2 percent of my body weight. However, ratios are just as misleading as absolute brain sizes, for a basic reason: Across species, brain size increases at a slower rate than body size. Across mammals, when body size triples, brain size barely doubles. Hence, other things being equal, the ratio between brain size and body size will gradually decline as overall body size increases. Small-bodied mammals generally have higher ratios than large-bodied mammals. Primitive mouse lemurs provide a useful example among primates. With a body weight of two ounces, a lesser mouse lemur is one of the smallest living primates. Yet an adult's brain makes up 3 percent of its body weight, leaving my 2 percent brain-to-body ratio in the dust.

Thus both absolute brain size and ratios between brain size and body size are misleading. Reliable comparisons of brain size between mammal species require special analyses that allow for the fact that the relationship between brain size and body size follows a decelerating curve. Only such analyses, first proposed by Dubois and embedded in a convincing overall framework by Jerison, permit meaningful interpretation of brain size in mammals.

If appropriate scaling analyses are used to take body size into account, relative brain sizes can be compared across species. Comfortingly, when body size is appropriately taken into account, we humans turn out to have the largest brain size among modern mammals. But what about that oft-repeated claim that primates have larger brains than other mammals? Well, this notion is still misleading, for two reasons. First, although *average* relative brain size is indeed larger for primates than for other mammal groups, individual primate species vary widely and overlap extensively with other mammals. In fact, a few lower primates have relative brain sizes below the overall average for mammals. Second, although humans do have the largest relative brain size found among mammals, most primates have distinctly smaller values. Relative brain size in humans is three times bigger than in great apes and twice as big as in the closest nonhuman primates, notably the New World capuchin monkeys. The yawning gap between human relative brain size and the highest values for nonhuman primates is occupied by dolphins and their relatives, which are members of an entirely different group of mammals. As regards relative brain size, some dolphins are remarkably, not to say uncomfortably, close to humans.

To sum up, it is simply wrong to imply that adult primates all have larger brains than other adult mammals. This is untrue in any sense—absolute, proportional, or appropriately scaled. It is crucial to recognize this because at birth primates do consistently have relatively larger brains than other mammals. While newborn primates have brains twice as big as in other newborn mammals at any given body size, this distinction between primates and nonprimates is far less obvious once adulthood is reached. Greater brain growth after birth allows some mammals to catch up with primates and even partially overtake them. For instance, once adult, many carnivores overlap with monkeys and apes in relative brain size. Carnivores have poorly developed, altricial newborns, so a large amount of brain growth occurs after birth. Mammals can evidently become large-brained adults in different ways. Although primates undoubtedly benefit from their head start at birth, that is not the end of the story.

Up to this point, brain size in humans and other primates has been discussed without reference to sex. Everything said thus far applies equally well to males and females. However, there exist some intriguing differences between males and females regarding the completed size of the brain. A developmental perspective is needed to tease apart some of the complex issues involved.

The first obvious feature, which has aroused some controversy, is a pronounced difference in brain size between adult men and women. On average, an adult woman has a brain size about 10 percent smaller than that of an adult man. This was first established soon after the Darwin/Wallace theory of evolution was announced. French anatomist Paul Broca, best known today for his discovery of Broca's area, a language area in the human brain, pioneered measurements of human brain size. His basic technique was to fill the braincases of human skulls with lead shot to measure their volumes. He noted the 10 percent average difference in braincase volume between men and women, triggering a debate that has continued ever since. Broca believed that brain volume is a meaningful indicator of intelligence, and much of his research was aimed in that direction.

There is a natural tendency to believe that brain size, as indicated by braincase volume, should tell us something useful about the owner's abilities. Partly for this reason, decades ago eminent men in Western society made arrangements for their brains to be removed and studied after death. It was simply assumed that outstanding men owed their abilities to larger brains and that the smaller average brain size in women reflected lesser intelligence. Paul Topinard—Paul Broca's leading disciple—published a paper early in 1882 in which he gave the following explanation for the observation that women have smaller brains: "A man, who must strive for two in the struggle for existence, bears all responsibilities and concerns for the future. He is constantly and actively challenged by the environment and by rival forces. So he needs a larger brain than the woman that he must protect and feed. She devotes herself to household tasks, and her role is to bring up the children, to love and to be passive." A few months later, he published another paper espousing the diametrically opposite view that the smaller size of the woman's brain is due to relative smallness of body size. He wrote, "I believe that I have been able to demonstrate that the sexes are equal as far as development of the brain is concerned. Indeed, it might even be claimed that women are more advanced in their evolution than men." Regretfully, I have not been able to find out what happened to change Topinard's mind over those few months. Perhaps he got married and had some sense beaten into him.

The oft-repeated conclusion that women are less intelligent because they have smaller brains is part of a larger picture in which human brain size is crudely equated with ability. Stephen Jay Gould effectively debunked this notion in his excellent book *The Mismeasure of Man*, but the pernicious tendency lingers on. In the first place, as in mammals generally, brain size is related to body size, so the fact that women generally have smaller bodies than men surely has something to do with their smaller brain sizes, as Gould correctly noted. Paul Harvey has commented that Gould—doubtless carried away by his enthusiasm for a just cause—removed the effect of body size twice over. Different body sizes, while they account for most of it, do not entirely explain the brain size difference between men and women.

Regardless of any sex difference, body size is a significant factor influencing brain size even among men. It has often been noted that the French novelist Anatole France, recipient of the Nobel Prize for literature in 1921, had the smallest brain on record, just over two pounds. Recently it has emerged that Albert Einstein also had a very small brain, barely 10 percent bigger than that of Anatole France. In this case, Einstein himself did not request removal of his brain. Acting on his own initiative, a pathologist friend took this step at the postmortem examination. As Michael Paterniti amusingly recounts in *Driving Mr. Albert: A Trip Across America with Einstein's Brain*, retrieval of the brain and accompanying documentation was a close call. But the bottom line is that Anatole France and Albert Einstein were both notably small men. Their small brains clearly reflect this basic fact, rather than anything about their intellectual ability.

Brain size in humans varies widely, partly because body size is so variable. Although average brain weight in modern humans is about three pounds, the overall range in perfectly normal people extends from two to four pounds. The largest normal human brains are about twice as big as the smallest. Wide variation is also evident when men are compared to women. Although a male brain is on average about 10 percent bigger than a female brain, overlap is extensive. Moreover, the range of variation is greater in men than in women.

The notion that men are more intelligent than women has proven to be strangely resilient in the face of much contradictory evidence. An authoritative 1995 paper by education experts Larry Hedges and Amy Nowell reviewed numerous previous studies of sex differences and variation in mental test scores. They showed that, overall, average values in intelligence test scores hardly differ between men and women, although men show greater variation. The authors noted, however, that men and women diverge somewhat for particular skills. Women tend to do better on tests that require writing skills, while men tend to do better on tests calling for mechanical skills. The crucial point is that despite the 10 percent difference in brain size, there is no overall difference in intelligence between men and women.

It is worth noting that the inventor of the intelligence quotient (IQ), Frenchman Alfred Binet, originally developed his test to identify students in need of special help with schoolwork. His noble aim was to ensure that disadvantaged children would receive educational support. Unfortunately, IQ testing is now widely used more to discriminate, and the notion of IQ has a negative connotation in many people's minds. Moreover, it is clear that scores achieved on IQ tests are influenced by cultural context and can be improved by training. There is no such thing as a culture-free IQ test that measures inborn ability.

In fact, IQ test scores are sometimes subject to a peculiar but little-advertised form of manipulation. When I was growing up in England, stepping up from primary to secondary school in the state system was based on a widely feared examination known as the "Eleven Plus," usually taken by pupils between the ages of eleven and twelve. IQ testing was used to assess mathematical ability and literacy. The outcome decided whether a schoolchild could go on to a university-track grammar school or was relegated to a lower-level secondary modern school. I still remember fretting about the approaching Eleven Plus examination and rejoicing when I passed. A few years ago, I found out that the examination results were systematically adjusted. At an age of eleven to twelve, girls consistently perform better than boys on IQ tests, so the results for girls were adjusted downward to ensure that roughly equal numbers of boys and girls went on to grammar schools. So much for superior male intelligence!

Development sheds additional light on differences between male and female brains. Although it is true that men have brains about 10 percent bigger than those of women, there is little sex difference at birth. Indeed, large samples are needed for a statistically significant result. The average brain weight of a newborn boy is just over 3 percent greater than the average brain weight of a newborn girl. Consequently, a boy's brain must grow more after birth to result in the 10 percent difference in brain weight between adult men and women. It has long been accepted that division of nerve cells (neurons), the basic components of the brain, stops about halfway through human pregnancy. What this means is that, unless the brain of a newborn boy has a higher density of neurons, it cannot have many more neurons than a girl's brain. Although recent evidence indicates that some division of neurons does occur during the second half of pregnancy and after birth, this makes little contribution to overall brain size. Then why is a man's brain 10 percent bigger than a woman's? Perhaps more connections are formed in the male brain, requiring more nerve fibers.

But there is a curious sex difference in human brain growth after birth. A girl's body generally grows faster than a boy's, and as a result, eleven-year-old girls are somewhat bigger than eleven-year-old boys. Yet most brain growth takes place early in life—by the age of seven years, the brain has almost reached adult size. What this means is that a seven-year-old boy has a brain that is some 10 percent bigger than the brain of a seven-year-old girl. This has a major consequences for energy requirements. Brains are energy guzzlers, so the almost adult-sized brain of a seven-year-old boy requires more resources than the almost adult-sized brain of a seven-year-old girl. However, there is another interesting implication. An eleven-year-old boy has an almost adult-sized brain that is 10 percent larger than the brain of a girl of the same age. Yet at the age of eleven girls do better than boys on IQ tests.

Let me propose an alternative, slightly outrageous explanation of the brain size difference between boys and girls. Brains are not exclusively made up of neurons and nerve fibers. They also contain glial cells, which apparently play no direct role in nervous processing. Instead, glial cells seem to have a support function, providing nutrients and a structural scaffold. They may also have a packing function, rather like polystyrene peanuts. Human males need bigger skulls because they have bigger jaws, teeth, and jaw muscles and generally have a bigger body. Perhaps the male brain merely contains more glial cells for packing and support, and not more neurons and connections between them. This directly leads to a prediction that neurons and nerve fibers should be less dense in male brains than in female brains. Present evidence is equivocal. Some authors have reported that at least some parts of male brains have lower neuron densities, while others deny that there is a difference. The jury is still out, but I see no evidence that there is any fundamental difference between male and female brains that translates into any significant difference in ability.

Of course, the most striking sex difference concerning the brain is the unique role the mother plays in its development. Throughout pregnancy and lactation, maternal resources are crucial for the offspring's developing brain. We in fact get our brains from our mothers. And that maternal contribution goes even further than I originally thought. A fascinating study by radiologist Angela Oatridge and colleagues used a magnetic resonance scanner to study brain dimensions in nine pregnant women. They discovered that a woman's brain decreases in size by an average of about 4 percent over the course of pregnancy and then takes about six months after birth to regain its previous size. Seemingly a human mother goes above and beyond the call of duty, cannibalizing her own brain to nourish her developing fetus.

CHAPTER 6

Feeding Babies: A Natural History of Breast-feeding

In the mid-1960s, I spent two years doing doctoral research with Konrad Lorenz at the Max Planck Institute in Seewiesen, Germany. My goal was to study the behavior of a breeding colony of treeshrews, squirrel-like inhabitants of Southeast Asian forests, in order to explore their evolutionary relationships. At the time, treeshrews were thought to be the most primitive living primates. So I figured that a detailed study of their behavior should yield clues to the habits of the common ancestor that gave rise to all living primates. Quite unexpectedly, I soon uncovered evidence that something was seriously wrong with the notion that treeshrews are close relatives of primates—and my discovery came from reproduction. Intensive mothering of well-developed infants is one of the most striking features that all primates share. Yet I found that a mother treeshrew makes a separate nursery nest where she gives birth to a litter of poorly developed babies, hairless and with their eyes and ears sealed. The most surprising thing is that she leaves her infants alone in their nursery, sleeps in her usual nest, and returns briefly to suckle them only once every forty-eight hours. During their monthlong stay in the nursery nest, the mother spends a total of just over an hour with her infants. Maternal behavior in treeshrews is the rock-bottom minimum found in mammals, contrasting starkly with the extensive mothering shown by all primates. This stunning discovery launched me on a lifetime's odyssey to find out how primates really did evolve.

As the previous chapter showed, mammal mothering is linked to brain size. Moreover, diligent parenting is often associated with strong social bonds. In all mammals, mothers invest in their infants' brains not only during pregnancy but also through suckling after birth. And wild-living mothers generally respond to newborn infants with appropriate care. Unlike infant treeshrews, our babies need intense, extended parental care. For a young couple returning home with their first newborn baby, parenthood can seem daunting. Advice is abundantly available, particularly through the Internet, but it covers a whole spectrum ranging from suggestions that a baby should be breast-fed for up to seven years to assurances that bottle-feeding is equally good and more convenient to boot.

Because culture has greatly influenced mothering in all modern human societies, it is not easy to decide what is "natural" for our own species. In the past, self-appointed experts concerned themselves with tidy rules, providing advice that largely ignored biology. Up to the middle of the nineteenth century, physicians and guidebooks rarely proposed routine schedules for breast-feeding. Then things changed quite rapidly as the industrial revolution gathered pace and working mothers became commonplace. By the beginning of the twentieth century, rigid breast-feeding schedules were the norm in the industrialized world. This new approach to breast-feeding, accompanied by reduction of the recommended duration from two years to one, was fostered by pioneering pediatricians Luther Emmett Holt in New York and Thomas Rotch in Boston. Parallel trends occurred in England, France, and Germany. Holt's best-selling book *The Care and Feeding of Children: A Catechism for the Use of Mothers and Children's Nurses* was first published in 1894. It eventually ran to a total of seventy-five editions and printings and was regarded as a definitive text on infant care until around 1940.

Holt's rule that an infant should be fed at three-hour intervals throughout most of the first year derives from his 1890 study of the stomachs of deceased babies. He measured each stomach's volume by clamping it at both ends and filling it with water. To calculate his feeding schedule, he then divided up the total daily milk intake into average stomachfuls. In a fascinating 1987 paper, American pediatrician Marshall Klaus reviewed the notion of rigid feeding schedules and lampooned Holt's rule as the "gas tank theory."

Fortunately, nature has left us with some clues to untangle the biology of breast-feeding from trend or hearsay. We can begin by looking at mothering in other primates, our zoological relatives. Just as the image of mother and baby is one of the most iconic in art—from ancient Egyptian images of Isis nursing the infant king Horus to Madonna and Child paintings by Leonardo, Dürer, and other masters—we often see appealing images of infant apes, monkeys, and even lemurs on the breast in zoos. For the basic activity of suckling, we can cast the net even wider, as all mammals show this behavior.

WE'RE MAMMALS. This isn't simply an arcane statement of biological classification. The message is far more profound: We have all the key biological features that set mammals apart from other animals. Two of them, hair and suckling, are obvious. What is not so obvious is that hair and suckling are linked. Let's consider hair first.

Mammals typically have hair, although some have reduced or lost most of it as a secondary development. Aquatic mammals such as dolphins and manatees, for instance, are often almost bare. Humans also count among the exceptions, famously headlined in Desmond Morris's 1967 best seller *The Naked Ape*.

Rare fossils with impressions of body outlines provide direct evidence that early mammals had hair. We know for sure that mammals had fur coats by 170 million years ago. And indirect evidence suggests that hair evolved earlier, more than 200 million years ago. Even before the first mammals evolved, advanced mammal-like reptiles had pits on their snouts for whiskers, which are specialized hairs.

As living mammals typically have hair, it's understandable that some classifications once used the name Pilosa (from the Latin *pilus*, "hair") for the entire group. But the other striking feature of mammals, suckling, is even more fundamental. It is truly universal, as exceptions don't exist. All female mammals produce milk to feed their infants by suckling.

Both hair and suckling are more basic and ancient than live birth. Among modern mammals, a few unusual inhabitants of the Australian region—platypuses and spiny anteaters—still lay eggs, but they have hair and suckle their infants. All modern mammals suckle infants, so it is highly likely that their common ancestor already did so. Live birth evolved later, after the monotremes had branched away, originating somewhere between the origin of all mammals and the common ancestor that gave rise to marsupials and placentals, at least 125 million years ago.

But hair and suckling are linked by more than an ancient origin. Different kinds of skin glands evolved along with hair. Biologists recognize three basic types: sweat-producing eccrine glands, scent-producing apocrine glands, and oil-producing sebaceous glands. The milk-producing glands of ancestral mammals likely evolved from sebaceous glands. Because their oily secretions help maintain fur condition, they are directly connected to hair follicles. Originally, milk-producing glands were also connected with hair follicles, providing a clue to their origin.

In ancestral mammals, skin glands producing moist secretions were gradually converted to mammary glands yielding milk with a mixture of nutrients and antibiotics. We tend to think that milk is simply a source of nourishment for the baby. And that can lead to the mistaken belief that artificial milks only need to deliver the appropriate nutrients. In fact, antibiotics contained in a mother's milk provide the baby's first line of defense against germs.

It is tempting to think that biological classifications now universally use the official label Mammalia, derived from the Latin *mamma*, "teat," because suckling is so fundamental. But did Mammalia really trump Pilosa in the naming game because biologists concluded that suckling is more important than a fur coat? Modern classifications began with Linnaeus, who chose the label Mammalia over Pilosa. However, science historian Londa Schiebinger discovered that Linnaeus actively campaigned to encourage Swedish women to breast-feed their babies, and indeed wrote a pamphlet about the topic. Thus we find that the emphasis on suckling rather than hair in his pioneering classification was more political than biological.

It is reasonable to infer that the common ancestors of all mammals suckled their offspring. Unfortunately, we cannot test this with fossil evidence as we can for hair. Even a complex feature such as suckling could have evolved separately in different lineages. Similar needs often lead to similar adaptations, so independent evolution of adaptations, or convergence, is quite common. When the ancestors of dolphins and whales returned to life in water, for instance, they developed a streamlined body form like that of fish. In a similar way, suckling might not have had a single origin, but how can we go about tracing its evolution? This is a dazzling case where genetic evidence—greatly strengthened by complete genome sequencing in diverse mammals—has come to the rescue within the last few years.

A prominent universal feature of mammal milk is the presence of special proteins known as caseins. Caseins are unique to mammals, and the genes that produce them are active only in mammary glands. Scientists have already sequenced complete genomes of an egg-laying monotreme (platypus), a marsupial (opossum), and five placental mammals (cow, dog, mice, rat, human). An evolutionary tree based on DNA sequences of casein genes reveals that there was only a single origin in the common ancestor that gave rise to monotremes, marsupials, and placentals. This finding resoundingly confirms the interpretation that suckling evolved only once in the earliest mammals.

Milk sugars provide additional evidence that suckling evolved in the earliest mammals. Comparisons show that milk-specific sugars were already present in the common ancestor of all mammals more than 200 million years ago. At that stage the sugars evidently were still diverse because different kinds dominate the milks of modern monotremes, marsupials, and placentals. However, in all placentals, including humans, lactose is the principal milk sugar, which means it was doubtless dominant in their common ancestor.

Lactose is a compound of two simple sugars, glucose and galactose. Compound sugars cannot pass through the wall of the intestine and must be digested to enter the bloodstream. By scientific convention, the name of any natural enzyme ends in "-ase," so the enzyme that splits lactose into glucose and galactose is known as lactase. In the offspring of all placental mammals, from birth onward, a single gene produces lactase exclusively in cells lining the intestine. The capacity to digest lactose typically disappears soon after weaning. Human children, for instance, commonly stop producing lactase by about five years of age.

Once lactase production has stopped, any lactose in the intestine will remain undigested and travel on to the colon intact. Bacteria inhabiting the colon rapidly adapt to digest available lactose. The resulting fermentation generates large quantities of a gas cocktail containing hydrogen, methane, and carbon dioxide. This explains why many adults who consume milk products are afflicted by digestive problems. As one reviewer wryly noted, this has "socially unacceptable consequences."

Because mammals do not normally consume milk after weaning, it is hardly surprising that many people can't tolerate lactose as adults. Yet as the tradition of consuming dairy products became routine in some societies, secondary adaptation occurred, allowing adults to digest lactose. As a result, there is marked variation between human populations. Compared to Asia, where lactose intolerance in adults is predominant, the level in Europe is quite low. The control mechanism of the lactase gene was modified so that its activity persists after weaning. In fact, this change occurred independently, with different genetic modifications, in central Europe and in northern Africa. Combined archaeological and genetic evidence suggests that lactose tolerance in adults originated about 7,500 years ago in dairy-farming communities in central Europe. In Africa, the picture is more complex: Four different genetic modifications permitting lactase persistence have already been identified, and there are probably more. Most of the changes originated separately in Africa thousands of years ago, but the typical European modification is also found there.

THE FUNDAMENTAL IMPORTANCE of suckling in mammals is indicated by the remarkably early appearance of the mammary apparatus during development. The basic framework of a developing infant's body slowly comes into being during the embryonic stage, which takes up the first two months of human pregnancy. By the beginning of the subsequent fetal stage, the main organs of the body are visible. In contrast, the mammary glands begin to develop in the middle of the embryonic stage, after the limb buds form and well before body organs are recognizable.

In all mammals, two parallel milk lines develop as thickened skin ridges on the embryo's underside, extending down left and right sides of the body from the armpit to the groin. In a human embryo, milk lines form in the fifth week after conception. Teats develop only later, usually in matching pairs, as outgrowths from the milk lines, which eventually disappear. In the human fetus, several teats start to develop along each milk line. Most of them usually disappear before birth, leaving one on each side. From birth onward, women—like monkeys and apes—typically have only a single pair of teats on the chest.

Because of developmental accidents, women sometimes have extra teats, although usually without any milk glands attached. Two, three, or even more pairs of teats are sometimes present, with or without local swelling. In 1886, gynecologist Franz Neugebauer reported a rare case of a woman with five pairs of teats. In addition to the usual teats on globular breasts, she had four accessory pairs: one in the armpits, two more above the breasts, and yet another below. When examined by Neugebauer, the woman was nursing an infant. In addition to the main milk supply from the breasts, some milk flowed from the teats in the armpits, and the other extra teats produced a little if squeezed.

Extra teats aroused no scientific interest five centuries ago, but Henry VIII reportedly had Anne Boleyn executed as a witch because she had a third nipple and perhaps even a breast along with it. Accessory teats were then seen as the devil's work, so this anomaly no doubt made a handy addition to the trumped-up charges of adultery and even incest that Anne Boleyn had to face.

Across mammals, the normal number of teats in adult females varies from one to ten pairs, but it is usually constant for a species. The number of pairs of teats corresponds to the average number of offspring at birth. In species such as humans, horses, and elephants, which typically have single births, one pair of teats is the norm. Species such as dogs and rats, which produce litters of offspring, generally have several pairs of teats. It is not clear why there are generally two teats for each offspring. In any event, pairs of teats also match the number of infants in most nonhuman primates. Species with a single pair of teats usually have single offspring, as is the case with apes and most monkeys. By contrast, certain lower primates that have two or three pairs of teats usually produce litters of two, three, or four offspring. For instance, the two-ounce mouse lemurs that I bred for my research had three pairs of teats and usually two or three tiny infants per litter. Thus a woman's single pair of teats tells us that humans are adapted for single births. On the other hand, the presence of several pairs of teats during early human development is a telling clue that distant mammalian ancestors had multiple offspring.

In *The Descent of Man,* Charles Darwin astutely noted a curious fact about nipples: Men, like adult males of many other mammal species, also have them. Men can also have extra nipples just like women. Moreover, as Darwin noted, "the mammary glands and nipples, as they exist in male mammals, can indeed hardly be called rudimentary; they are merely not full developed, and not functionally active."

Intriguing exceptions exist among mammals. Male marsupials, mice, rats, and horses all lack teats. Yet most male mammals have them, including guinea pigs, carnivores, bats, and primates. Human infants of both sexes already have teats at birth, although they are nonfunctional until much later and then only in women. When born, boys and girls both have a small swelling beneath each nipple and may produce "witches' milk." This is a short-lived aftereffect of hormones produced by the placenta, but it does show that milk can flow from male nipples.

Long before birth, in mammals generally, the milk lines develop in both male and female embryos. Why does this happen in both sexes, and why do teats of most male mammals persist into adulthood? In fact, many scientists wonder why men do not suckle along with women. At first sight, it seems like male suckling would provide a welcome backup, enhancing infant survival.

As females alone bear the burden of pregnancy, it may simply seem par for the course that males leave females to carry the burden of nursing as well. But other forms of paternal care, such as infant carrying, have evolved in mammals. Why, then, don't males suckle? Evolutionary biologists have concluded that males should only help females raise offspring if paternity is certain. Genetically speaking, any male who helps rear offspring that are not his own will be at an evolutionary disadvantage compared to a male who provides no help at all. Carrying the burden of suckling alone may be the price a female mammal pays for even thinking about infidelity.

Early development of milk lines in males and females raises the intriguing possibility that long ago in mammal evolution both sexes suckled, but males then gave it up. In *The Descent of Man*, Darwin actually stated that at one stage of mammal evolution "both sexes yielded milk, and thus nourished their young; and in the case of marsupials, that both sexes carried their young in marsupial sacks." However, he did not convincingly explain why males suckled early in evolution and then stopped; he simply suggested that smaller litters possibly made male help redundant. We also need to explain why male marsupials, mice, rats, and horses do not have teats. For the time being, we have to accept that we really have no good explanation for male teats. Still, those thousands of men who suffer from "jogger's nipple" surely deserve an answer.

It was long thought that male mammals never produce milk under natural conditions. Then biologist Charles Francis, bat expert Thomas Kunz, and two colleagues reported that male Dayak fruit bats produce milk. During a field study in Malaysia, the team managed to express milk from the teats of all thirteen mature males captured with mist nets. Moreover, microscopic examination of a few male specimens confirmed that mammary tissue was actively secreting milk. It was later reported that males of another fruit bat, the masked flying fox of Papua New Guinea, also produce milk. But in both bat species males produce only tiny amounts of milk compared to females. Moreover, in Dayak fruit bats male nipples are smaller and less hardened than those of females. This suggests that little, if any, suckling occurs. In any event, suckling by male mammals is exceedingly rare, if it exists at all, and that task generally falls to females.

NOZZLE-LIKE TEATS are needed for suckling. The globular breasts of women, plumped up with fat deposits, are an optional extra. Yet we talk of breast-feeding, not nipple-feeding. Permanent swelling of the breast bearing the nipple is not essential either for milk production or for suckling. So why do women have swollen breasts?

Contrary to expectation, milk is not exclusively or even mainly produced from fat deposits in a woman's breasts. In fact, most mammals show little external swelling around the teat even while suckling. In others, swelling develops only when the first infant is suckled and may subside afterward. Female elephants, for example, develop conspicuously swollen breasts when suckling, but those usually regress after weaning. In various primates—including our closest relatives, the great apes—the chest area similarly remains flat until the first birth. Visible swelling is present during suckling and sometimes persists after weaning. Swelling of the chest area commonly persists in older female chimpanzees and gorillas after they have reared several offspring.

Following Desmond Morris, popular authors have often claimed that women are unique among mammals in two respects. First, permanent breast development occurs as one of the changes leading up to puberty, long before the first infant is suckled. Second, no other female mammal has such big swellings compared to the rest of the body. But such claims of human "uniqueness" require careful scrutiny. Rapid development of the mammary glands starts well before puberty in many mammals, including the humble rat. Moreover, the udders of ruminants are swollen by puberty. Of course, domestication has greatly increased their size, particularly in dairy cows, but even in wild-living ruminants udders are noticeably swollen.

As permanent swelling of the human breast around puberty is not needed for milk production or suckling, it likely has some other function. Although we take women's globular breasts for granted, it is puzzling why they look they way they do. Inarguably they attract attention. A 2010 study by Barnaby Dixson and colleagues used an eye-tracking device to detect where a man's gaze first falls when looking at an image of a woman's body. In less than a fifth of a second, almost half the men tested looked at the woman's breasts first, while one in three looked at her waist and one in seven looked at the pubic area or thighs. Just one in sixteen men looked at the woman's face first. Keep in mind that only cross-cultural studies can reveal responses that are consistent regardless of social norms; in societies where women traditionally bare their breasts, men's responses may well be quite different.

The enigma of swollen human breasts persists, despite many attempts to explain it. In 1987, wildlife biologist Tim Caro wrote an article slyly titled "Human Breasts, Unsupported Hypotheses Reviewed," listing no fewer than seven different possible explanations for the evolution of permanently swollen human breasts. One suggestion, already broached by Darwin, is that they evolved as signals to stimulate men's sexual behavior. Breasts are explicitly linked to sexual behavior in many, if not all, human societies, and sexual attractiveness would at least explain why human breasts swell during puberty rather than when the first baby is suckled.

Various other intriguing possibilities have been proposed to explain permanent breast enlargement in women. It has been suggested, for instance, that continuously swollen breasts enabled females to hide their reproductive condition from males, thus concealing paternity. An alternative proposal was that pendulous breasts allow an infant to suck while a naked mother carries it on her hip. However, this cannot explain the development of swollen breasts at puberty rather than just before the first birth. It has also been suggested that breast size could advertise potential fertility or suckling capacity. But, as already noted, milk production is not specifically linked to breast fat deposits. Nevertheless, breast size might provide a guide to the total amount of stored fat, perhaps indicating a woman's ability to withstand food shortage. The problem with this is that the breasts of an average woman store only 4 percent of total body fat.

A major drawback with all proposed explanations for swollen human breasts is that none has been properly tested. The little we do know indicates no clear link between breast size and breeding performance. Interestingly, medical studies have shown that the capacity for milk production soon after birth is associated not with the customary size of the breast, small or large, but with its *increase* in size during the last six months of pregnancy. So it is the capacity to enlarge the breasts, not their starting size, that seems to indicate milk output.

Whatever the explanation for permanently enlarged breasts, many women in Western society—encouraged by plastic surgeons—have bought into the belief that artificial enlargement enhances sexual attractiveness. Thanks to the miracles of modern medicine, any woman can undergo surgical manipulation to enhance her breast size.

Another unusual, perhaps even unique feature of the human breast is a prominent area of pigmented skin, the areola mammae, surrounding the teat. The areola bears the main impact of suckling. The tip of the teat itself is surrounded by a circular array of fifteen to twenty duct openings that release milk during suckling. The areola also has several small openings that may project as little bumps above the surface. These are outlets of Montgomery's glands, special sebaceous glands that lubricate the area around the nipple, protecting it and facilitating suckling. Coloration of the areola varies from light pink to dark brown according to general skin pigmentation. It also varies over time along with hormonal changes across the menstrual cycle, pregnancy, and the overall life span. Everything indicates that the primary function of nipples is for suckling, and that is where attention should be focused.

Beginning with birth and ending with weaning, every mammal mother suckles her offspring for a certain amount of time, called the lactation period. Culture has greatly influenced mothering in all modern human societies. Consequently, it is no easy task to decide what is "natural" for our own species. As we seek clues, it is once again very useful to survey primates—and other mammals as well—to identify general principles. In many species, suckling duration is remarkably constant. In others, particularly in large-bodied mammals with single infants, it is quite variable. A mother house mouse typically suckles her pups for twenty-two days, a rat for thirty-one days, and a treeshrew for thirty-five days. Their lactation periods show remarkably little variation.

Mice, rats, treeshrews, and other similar mammals have a primitive breeding pattern, with short pregnancies and altricial offspring. Suckling stops sharply at a standard interval after birth and there is an abrupt shift to solid foods. Primates, by contrast, give birth to precocial offspring after long pregnancies. In many cases, suckling periods are variable and associated with a gradual transfer to solid foods. Among primates, suckling duration varies according to species, from a fairly constant forty-five days in a two-ounce mouse lemur to a variable period averaging around five years in a ninety-pound orangutan. The maximum duration of over seven years reported for an orangutan may be the longest recorded among mammals.

Like many other features, suckling durations are scaled to body size across mammals: the larger the mammal, the longer the lactation period. But, even compared to mammals of the same body size, primates generally suckle infants for a comparatively long time. There is also a grade shift among primates—at any given body size, monkeys and apes tend to suckle longer than lower primates. With all of this variation, it is difficult to decide on an average weaning age for which we humans are biologically adapted. Social norms governing weaning practices differ widely between human societies and change over time. Practices range from nursing for up to six years or more to not breast-feeding at all, resorting either to bottle-feeding or using wet nurses. The age at which lactase production usually stops, around five years, gives us one clue to a natural weaning age for humans. Unfortunately, the timing is too variable to provide more than a hint.

Anthropologist Katherine Dettwyler, who began her career with an award-winning field study of mothering in Mali and went on to become a prominent advocate of breast-feeding, became keenly interested in suckling time. She set out to estimate a natural human weaning age from various angles. One approach is to take a hint from modern human societies with a gathering-and-hunting lifestyle. They would give us an average weaning age for conditions closer to those that existed for at least 99 percent of human evolution. Gathering-and-hunting societies have no domesticated mammals as an alternative milk source, so cultural practices have less impact on weaning age. But we mustn't forget that early supplementary feeding can affect breast-feeding duration in all contemporary human societies.

It turns out that breast-feeding generally lasts around three years in gathering-and-hunting societies. Anthropologists Melvin Konner and Carol Worthman reported that children are weaned at an average age of three and a half years among the !Kung, gatherer-hunters of Botswana and Namibia. Konner and Worthman's pioneering two-year study became a textbook example of fieldwork in human biology. American Daniel Sellen is another anthropologist who has devoted his career to exploring the evolutionary background to human mothering. His work revealed that the long duration of suckling reported for the !Kung is typical of nonindustrialized societies generally. In a 2001 review of more than a hundred nonindustrialized societies, Sellen showed that weaning occurred at an average age of twenty-nine months, with an overall range between one year and five and a half years.

A second approach adopted by Dettwyler and others in seeking a biological clue to human weaning age is examining the consistent overall relationship between suckling duration and body weight for monkeys and apes. The average value expected for a woman can then be calculated from that relationship. Weaning age estimated in this way is also close to three years. Incidentally, Dettwyler practices what she preaches, having breast-fed her daughter, Miranda, until the child was four years old.

An estimated natural human breast-feeding period of about three years probably seems surprisingly long. Even so, it may be on the short side compared to our primate cousins. It is below the averages for wild-living great apes: four and a half years for common chimpanzees, three and a half years for gorillas, and five years for orangutans. And adult female chimpanzees weigh in at about ninety pounds, markedly less than the average woman. Thus weaning in chimpanzees should be expected to occur earlier than in humans, not later. In fact, apes tend to have somewhat later weaning ages, relative to body size, than monkeys. Because of this difference, the natural suckling duration of three years inferred for humans from a combined examination of monkeys and apes may be too low.

On the other hand, perhaps a special adaptation occurred after our early ancestors diverged from chimpanzees. For instance, adaptation for a high-energy diet during human evolution could have permitted nutrient-rich supplementary feeding of babies. Some experts believe that this might have allowed earlier weaning. Nevertheless, the balance of evidence, both from other primates and from human gathering-and-hunting societies, indicates that a natural weaning age in humans would be *at least* three years. As humans diverged from great apes in evolution, they probably maintained a basic pattern of late weaning that is still evident in modern gatherers-and-hunters living close to nature.

A natural weaning age of at least three years might come as a shock to women who are used to nursing their babies for three or maybe six months. It may temper the shock to know that the figure of three years is for total duration of breast-feeding. Cross-cultural research by Daniel Sellen and others indicates that exclusive breast-feeding generally lasts six months to a year. For the rest of the time until weaning the infant receives supplementary foods as well as breast milk. In 2005, the Section on Breast-feeding of the American Academy of Pediatrics recommended that, wherever possible, an infant should be exclusively breast-fed for six months and weaned at a year. Both the World Health Organization and the United Nations Children's Fund (UNICEF) have also advocated six months of exclusive breast-feeding but now recommend weaning at two years. So we are inching our way back to the timing that biological and anthropological comparisons suggest.

ADDITIONAL CLUES to the biological roots of human mothering, including mother-infant contact, come from a somewhat unexpected source: milk composition. Having started out as converted skin glands, mammary glands of modern mammals now produce highly specialized milk. In each species, milk is adapted to meet the specific needs of developing offspring. It also reflects the mother's typical suckling pattern. In fact, there are fine-tuned changes between birth and weaning, with daily production first rising and then declining to zero at weaning. For a human mother, it takes about five days for milk production to get properly under way. The yield then climbs from about a pint a day to two pints or more at peak lactation. Milk composition also changes to some extent over time. Apart from such fine-tuning, it shows relative constancy within a species.

Milk is a complex cocktail with many ingredients, and it is an important source of water for the suckled infant as well. Some ingredients are specially adapted, while others are less important or mere by-products. In simple nutritional terms, milk has three main components: fats, sugars, and proteins. Crudely speaking, proteins such as casein are major growth components, sugars such as lactose are a short-term energy source, and fats are for long-term energy storage. Fats also contribute to growth, particularly in cell membranes.

Comparisons of fats, sugars, and proteins in milks of different mammals reveal several important principles. For one, species that give birth to poorly developed altricial infants have milk that is quite rich in protein so that their young can grow rapidly. This pattern is typically the case with carnivores, insectivores, rabbits, rodents, and treeshrews. By contrast, milk protein content is low in species with well-developed precocial infants. This is true of elephants, hoofed mammals, and primates, notably humans. Average protein level in primate milks is below 2.5 percent, reflecting the fact that primate offspring grow slowly even compared to other precocial mammals. Human milk has the lowest protein concentration found in any primate, at just 1 percent.

Surprisingly, it is fat content that tells us the most about the typical suckling pattern of any species. Biologist Devorah Ben Shaul identified a crucial difference among mammals during a twelve-year project in which she analyzed milk composition as an aid to hand-rearing baby mammals in zoos. Her initial expectation was that milk composition would reflect evolutionary relationships between species. However, she eventually realized that nursing behavior and ecology are the main influences. Ben Shaul recognized a primary distinction between species that suckle on schedule, with the mother controlling intervals between suckling events, and those that suckle on demand, with the infant deciding the intervals. The general rule is that suckling on schedule is typical of mammal species in which mother and infant are separated for extended periods, while suckling on demand characterizes species in which they continuously remain close together.

Suckling patterns differ greatly between different mammal groups. In contrast to primates, mothers in some mammal groups have remarkably little contact with their offspring between birth and weaning. Treeshrews provide an extreme case of suckling on schedule in which the mother alone controls the pattern and timing. She suckles her infants only once every two days with a milk that is similar in consistency to whipping cream. The only other altricial mammals that come anywhere close to tree-shrews are wild rabbits. A doe keeps her pups in a separate chamber in the burrow, feeding them just once every day. Treeshrews and rabbits together occupy the lower end of a spectrum of mother-infant contact that extends up to the intensive parental care of primates.

The typical primate pattern of suckling on demand differs starkly from the extreme suckling on schedule shown by treeshrews and rabbits. In primates, the infant decides on the timing of suckling bouts. In most species, the baby rides directly on the mother's fur, so it can simply move to a teat whenever it wants a drink. Suckling on demand surely has been a key feature of mothering in primates since their common ancestor some 80 million years ago.

Seen in this light, various past prescriptions for tightly regulated breast-feeding of human infants, most explicitly in child care manuals in vogue in the United States and Europe between 1850 and 1940, are simply shocking. Any consideration of natural biological patterns to promote the welfare of both mothers and infants was swept aside by a misguided concern for rigid schedules.

Ben Shaul's findings regarding the relationship between mothering style and milk composition are fundamental. Suckling on schedule is typical for nest-living mammals with poorly developed, altricial offspring, such as treeshrews. Altricial nestlings are commonly left alone for long periods, and fat-rich milk provides a slow-release energy source that allows the nestlings to stay warm. Furthermore, altricial nestlings are not very active. They have little need for a rapid-release energy source, so milk sugar content is usually low. By contrast, suckling on demand prevails among precocial mammals, which rarely use nests. Well-developed precocial infants are often quite mobile and able to follow the mother around from birth onward. Elephants, dolphins, and most hoofed mammals such as cows and horses are prominent examples. Infants need rapidly released energy, so the milk tends to be rich in sugar but poor in fat.

Primates are unusual compared to other precocial mammals because their infants don't move around independently until long after birth. Close physical contact is ensured by carriage or nest sharing, with the benefit that the mother's body warms the infant. This feature greatly reduces the infant's need for energy to control its own body temperature. The upshot is that milk fat content is particularly low in primates generally, averaging less than 4 percent. Short intervals between bouts are typical for precocial mammals that suckle on demand, and frequent suckling bouts are certainly a prominent feature of primates. So overall milk concentration is generally low in precocial mammals. But, because infants are usually active from birth onward, the milk sugar level tends to be higher than in altricial mammals. Fitting this general pattern, primate milks contain relatively little protein and fat but have high sugar levels, averaging almost 7 percent.

Human milk composition closely resembles that of other primates. This is evidence that we are biologically adapted for the same basic pattern of close mother-infant contact associated with suckling on demand. That said, our babies do show at least one special feature when it comes to nursing. This is the "burst-pause pattern" noted by psychologist Kenneth Kaye in a paper on sucking by young infants. From birth onward, infants typically show a sucking rhythm that is standard for our species. Bursts of about twenty sucks are followed by distinct pauses with the nipple still held in the infant's mouth. This burst-pause sucking pattern seems to be unique to humans. As Kaye noted, it strengthens communication between the mother and the infant on the breast. This behavior suggests that the fundamental primate pattern of close mother-infant contact has been further enhanced, not reduced, in humans.

So far I have focused on the nutritional content of milk, but the purpose of breast milk is not simply to nourish babies. It has additional benefits. For instance, a mammal mother also provides her offspring with a cocktail of antibiotic ingredients. She provides her infant with temporary, passive protection against microbes while its own active defense mechanisms are developing. Protection against infection, in fact, may have been one of the earliest functions of suckling. Pediatrician Armond Goldman has noted that in mammals the oily secretions of sebaceous glands (the likely precursors of mammary glands) contain immune factors similar to those present in milk. Nutritionist Bo Lönnerdal reviewed several significant features of human milk and found that most of the specifically active ingredients are proteins, although there are various immune factors including antibodies and immune cells.

There are also beneficial bacteria that take up residence in the digestive tract. Babies are sterile at birth and have to pick up the bacteria they need.

The natural source is the breast-feeding mother. Thus harmless bacteria living in the gut also differ between breast-fed and bottle-fed infants, although suitable supplements can be added to milk formula to solve this problem.

In a 1995 article, Jack Newman, a pediatrician who founded the influential breast-feeding clinic at the Hospital for Sick Children in Toronto, reviewed the protective elements that human milk provides against noxious microbes. Newman noted that in several countries mothers directly use their breast milk to treat eye infections in infants. In fact, a child's own immune response does not reach full strength until it is about five years old, so the protection provided by mother's milk is sorely needed. Doctors have long recognized that breast-fed infants contract fewer infections. They suffer less than bottle-fed infants from meningitis or infections of the gut, ear, respiratory system, and urinary tract. That difference applies even when infants are fed with milk formula that has been sterilized.

All human babies receive some protection even before birth. Antibodies pass across the placenta to the fetus during pregnancy, and they continue to circulate in the infant's blood for weeks or even months after birth. From birth onward, breast-fed infants receive extra protection from antibodies, other proteins, and immune cells in human milk. Some proteins bind to microbes in the gut cavity, stopping them from passing through the gut wall. Others decrease the supply of certain minerals and vitamins that noxious bacteria need to survive in the gut. For instance, a special binding protein reduces the availability of vitamin B_{12}, while lactoferrin captures iron. Bifidus factor actively promotes growth of beneficial bacteria in the infant's gut.

Over and above the basic types of antibodies, human milk contains numerous immune cells, including some that attack microbes directly. The most abundant type of antibody in human milk is secretory IgA, which includes a component that shields it from digestion in the infant's gut. Bottle-fed infants have little to combat noxious microbes until they begin producing their own secretory IgA, usually some weeks or months after birth. In Newman's concluding words: "Breast milk is truly a fascinating fluid that supplies infants with far more than nutrition. It protects them against infection until they can protect themselves."

Around the time of birth, the mother produces a special kind of yellowish, low-fat milk called colostrum. This is a widespread, probably universal feature of mammals. Its primary, vital function is to transfer immunity from mother to offspring right after birth. Immune cells and the antiviral agent interferon are concentrated in colostrum, and it also includes growth factors that promote development of the infant's digestive tract. Thus it is especially important for newborn infants, including human babies, to receive the first batch of milk that the mother produces. Incidentally, cow colostrum has been used as a supplement to speed recovery of athletes after injury and (regrettably) to improve performance. Yet Western society did not recognize the significance of colostrum for the health of human babies until the latter part of the seventeenth century. Previously colostrum was widely believed to be harmful. This peculiar view dates back at least as far as claims made by the second-century Greek physician Soranus of Ephesus, was apparently widespread among preindustrial societies, and persisted in medieval Europe. This offers a striking example of the way in which cultural norms sometimes clash directly with biological reality.

SOME TWENTY-FIVE YEARS AGO, I gave a seminar talk on primate brain evolution at the University of Cambridge and learned the hard way not to exaggerate the negative effects of bottle-feeding. Unique demands of brain growth after birth, I argued, presumably require special adaptation of human milk. I rashly followed this entirely reasonable conclusion with an off-the-cuff comment about deficits in brain development due to bottle-feeding with artificial milk. After my presentation, seminar host Nicholas Davies—a truly gifted and eminent biologist—stood up to open the discussion. He immediately took the wind out of my sails by starting his thanks with the words "Speaking as a bottle-fed baby . . ."

So let it be stressed that human babies can thrive when reared with milk that differs in many ways from that of natural mothers. Drawbacks of bottle-feeding are statistical, not all-pervading. It is also important not to forget the inevitable time lag in studies comparing breast-feeding with bottle-feeding. Many reported results indicate effects of artificial milks used decades ago. In the meantime, step-by-step changes have doubtless improved bottle-feeding.

The practice of bottle-feeding started with using milk from domestic mammals, so it has a relatively short history. Domestication of mammals, which arose independently at a dozen different locations around the world, dates back no further than about 10,000 years. It probably did not take too long after the first steps in domestication of hoofed mammals for people to start feeding infants with animal milks. In fact, pottery vessels used in infant nursing are among the oldest containers we have uncovered, dating back almost 4,000 years.

Anthropologist Tosha Dupras has studied infant feeding and weaning practices during the Roman period in Egypt some 2,000 years ago, at the Dakhleh oasis. Documentary and other direct evidence for the Roman period in Egypt is limited, so Dupras and her colleagues analyzed stable isotopes in skeletons instead. Nitrogen and carbon isotopes capture distinct patterns during breast-feeding and at weaning. The study revealed that Egyptian mothers at this site introduced supplementary foods at around six months of age and completed weaning by three years of age. Investigation of isotopes in animal and plant remains from an ancient village nearby yielded valuable additional information. After the age of about six months, infants were fed with milk from goats or cows.

Additional supporting evidence from ancient times is provided by documentary sources for the Pharaonic period in Egypt (2686–332 BC), indicating breast-feeding of infants up to an age of three years. During this period, milk from other mammals was sometimes given as a supplement to older infants. Even earlier evidence is available from studies of nitrogen isotopes in skeletons of infants and children from two Neolithic sites in Anatolia, Turkey, dating back around 10,000 years. Archaeologist Jessica Pearson uses isotope analysis to glean clues about foods eaten by past populations and the relationship between diet and health. She also studies skeletons from archaeological sites to seek features that indicate past activities. Her team reported that in their study populations, exclusive breast-feeding lasted one to two years and weaning occurred between two and three years after birth. Both Anatolian communities were on the cusp of the shift from gathering and hunting to agriculture, harvesting a few domesticated plants and living with some not-yet-domesticated animals.

In modern societies, by contrast, bottle-feeding has become so common that the biological origins of human suckling have faded into obscurity. For many women, especially those who work, not only is it more convenient to feed the infant with artificial milk from a bottle, but it is nearly impossible to do otherwise. Howeve, these are relatively recent developments in a large-scale drift from breast-feeding to bottle-feeding. Bottle-feeding surely began soon after the origins of domestication, and various mammals have been used as sources of milk. Perhaps bottle-feeding, like wet nursing, first originated as an emergency, lifesaving procedure for motherless infants.

Breast-feeding has many advantages, but it does not always work out the way the mother would wish. Moreover, there are serious medical cases where breast-feeding may not be an option. For instance, human immunodeficiency virus (HIV) can be transmitted in breast milk, so bottle-feeding may be recommended for this reason. Whatever the cause may be, many babies today must be bottle-fed. Rather than agonizing over this necessity, we ought to use evolutionary biology to understand how best to supplement these children's nutritional and antibiotic needs so that any artificial milk can be as "natural" as possible.

AN IMPORTANT QUESTION to ask when considering whether to bottle-feed an infant is whether artificial milk provides the ingredients a baby needs. Examining the composition of human milk seems an obvious place to start. But milk contains a multitude of substances, including no fewer than two hundred different kinds of fatty acids—only a few of which may be specifically adapted to meet the infant's needs. It would in any case be virtually impossible to produce a precise chemical replica of human milk. Luckily for us, we have a valuable shortcut in our knowledge of suckling in mammals in general and primates in particular. Broad comparisons can help us focus on ingredients that are especially important for infant development. Extraordinary features of the human infant that might make precise demands on milk ingredients must also be considered. A prominent example here is our large brain size at birth, followed by unusually rapid growth for a whole year. This pattern is unique among mammals, so milk ingredients directly connected with brain development are likely to be especially important.

Yet, surprisingly, biological comparisons played little part in the history of bottle-feeding. Artificial milks were developed mainly by trial and error. At first, human infants were fed with milk from readily available domesticated mammals. After that, the foreign milk was gradually altered in various ways to address specific problems. Most traditional sources of milk for human infants are even-toed hoofed mammals—mainly cows but also buffalos, goats, sheep, camels, and llamas. All of these domesticated mammals happen to have well-developed precocial infants, like primates, so the basic composition of their milks is broadly similar. This approximate correspondence owes more to luck than to judgment, because large-bodied mammals with precocial infants are better prospects for domestication. Small-bodied mammals with altricial infants—such as cats, dogs, and rats—are pets or pests rather than beasts of burden and food sources.

As discussed, primates are unusual among precocial mammals. For instance, primate milk is less concentrated than that of even-toed hoofed mammals. It was likely quickly noticed that milk from any domestic ruminant had to be diluted so that it was not too rich for a human baby. Until a human infant is about a year old, raw milk from a cow or any other ruminant is considered unsafe for consumption, as it can lead to bleeding from the intestine. It was also eventually recognized that sugar should be added to diluted cow's milk to make it more similar to human milk. Interestingly, in overall composition horse milk seems to be closer to human milk than that of any even-toed hoofed mammal such as a cow. However, milking a horse regularly is far more challenging than milking cows, goats, or sheep—a mare's teats are located right between her muscular hind legs.

Over the years, manufacturers in industrialized nations have repeatedly tweaked the milk formula used for bottle-feeding. Still, the bottom line is that countless people today continue to feed babies with an adulterated form of cow's milk. Hoofed mammals and humans last shared a common ancestor about 100 million years ago. In view of that huge evolutionary gulf, it is remarkable how well our babies do on this substitute diet. Indeed, it might seem at first sight that artificial milks are just as good as human milk. The digestive system of the human infant has an amazing degree of built-in flexibility. It can extract what it needs from surrogate milk with ingredients that reflect a judicious balance between adequate nutrition and healthy profit margins.

One of the reasons for this flexibility may be that, as biologist Caroline Pond noted in her fascinating 1998 book *The Fats of Life*, fats in mammal milks are largely used as fuel, not for growth. The specific array of different fats in milk may not be very important for the suckled infant. But how far does the apparent adaptability of human infants go? It is essential to know whether bottle-fed babies really do as well as babies raised on the breast.

Obviously, any difference between bottle-fed and breast-fed babies could not pose a major threat to life. Any serious health risk surely would have aroused attention and appropriate counteraction long ago. On average, bottle-fed babies clearly develop well enough that their health is not seriously compromised. The absence of class action lawsuits in the United States speaks for itself. On the other hand, the wide range of natural variation in infant development could easily mask relatively subtle differences. Only careful examination and meticulous statistical testing can reveal minor deficits associated with bottle-feeding.

One suggested drawback is that bottle-fed infants may have difficulty regulating intake. This could result in overfeeding, increasing the risk of childhood obesity. We shall return to this further on. Bottle-feeding can also introduce additional problems through use of contaminated water, overdilution, and unsuitable weaning practices. It might be thought that contamination of bottle-fed milk is a problem confined to third world countries, but it has recently emerged that some baby bottles in industrialized countries contain the potentially harmful substance bisphenol A (BPA). I highlighted this insidious chemical in Chapter 1 because of its connection with decreased sperm counts. Exposure to BPA may impair development of the brain and thyroid gland as well, so even the bottles used to feed artificial milk can be dangerous. Problems of contamination may partly account for epidemiological studies revealing that bottle-fed infants grow up into adults who suffer higher incidences of diabetes, cancer, and cardiovascular disease.

In the search for key ingredients of human milk, much attention has focused on a particular class of complex but vital fats: long-chain polyunsaturated fatty acids (LCPUFAs). Simply stated, unsaturated fatty acids can add chemical links, whereas saturated fatty acids cannot. The structural difference has a practical implication: polyunsaturated fatty acids have a lower melting point and remain liquid at body temperature. Probably in connection with this, LCPUFAs are important structural components of cell membranes.

LCPUFAs are especially well represented in nerve cells. Optimal development and function of the nervous system requires an adequate supply. Nutritional researchers such as Susan Carlson, Michael Crawford, and Stephen Cunnane have emphasized the importance of these fatty acids for normal brain development during pregnancy and breast-feeding. Two important examples are arachidonic acid (AA) and docosahexaenoic acid

(DHA). AA and DHA are prominent ingredients of nutritional supplements containing omega-6 and omega-3 fatty acids, respectively. They are key components of nerve cells, and DHA is also crucial for light-sensitive cells in the retina of the eye.

It is unclear whether a growing human baby can manufacture all the LCPUFAs it needs or whether the mother must supply them. In view of the unique demands of human brain development after birth, these fatty acids are probably crucial ingredients of human milk. It is certainly true that they are well represented in human milk specifically and in primate milk in general. Perhaps the special needs of the growing human brain are met by simply providing enough milk to meet the overall need for LCPUFAs. But cow's milk contains only trace amounts.

It is also likely that LCPUFAs stored during fetal development contribute to human brain growth after birth. As indicated in the previous chapter, Stephen Cunnane and Michael Crawford have suggested a connection with the unusual plumpness of newborn human infants: Stored fat may provide LCPUFAs to support brain development. It is also possible that early provision of suitable complementary foods rich in these fatty acids could boost availability for brain development in human infants.

Because cow's milk has only trace amounts of LCPUFAs, there is a possibility that bottle-feeding could lead to deficient development of an infant's nervous system. It is known that blood concentrations of LCPUFAs are higher in breast-fed infants than in bottle-fed ones. Circumstantial evidence also indicates that development of the nervous system may be adversely affected in bottle-fed infants. Results reported for infants born after full-term pregnancies have been mixed, but for preterm infants born after an unusually short pregnancy there is clear evidence that a shortage of LCPUFAs in artificial milk is detrimental. The jury is out regarding a general need for these fatty acids in bottle-feeding, but it is pretty clear that milk given to preemies should contain enough of these important fatty acids.

Fat is stored by the fetus only during the last three months of pregnancy. Thus infants born well before the due date lack the customary plumpness and have unusually limited fat reserves; their need for LCPUFAs supplied in milk is hence much greater. Because of mounting evidence that LCPUFAs in milk may be important for normal development of the nervous system, AA and DHA have been gradually added to artificial milks in various countries. In 2002, the U.S. Food and Drug Administration belatedly approved addition of AA and DHA to milk formula. Yet artificial milk enhanced in this way was not intended for preterm infants, despite this group having the greatest need for supplementation with LCPUFAs. The basic problem has been that evidence indicating the vital importance of AA and DHA in human milk has generally been indirect. That evidence, however, is a smoking gun. Clearly this is an urgent topic for targeted medical investigation.

STARTING IN THE 1970S, investigators began to reveal developmental shortcomings in bottle-fed infants. One landmark study, conducted by medical researcher Bryan Rodgers, examined children included in the 1946 birth cohort monitored by the National Survey of Health and Development in the United Kingdom. Attainment tests were conducted with 2,000 children when they reached eight and fifteen years of age. After the researchers took differences in family background into account, they discovered that children who had been entirely bottle-fed in infancy scored significantly lower than those who had been entirely breast-fed, although the difference was only a few points.

Several subsequent studies have yielded similar results, confirming that bottle-fed children *on average* achieve lower scores on intelligence tests and are more likely to suffer learning deficits. Although the effects are small, they are reportedly statistically significant.

All surveys of this kind face the problem of confounding factors, which I illustrated with the stork-and-baby example in Chapter 2. To take a real case in the realm of breast-feeding, we know that mental development is affected by economic circumstances. On average, babies of higher-income women perform better on mental tests than their lower-income counterparts. However, there is also evidence that well-off women are more likely to breast-feed. Because of this, breast-feeding may seem to be associated with results of mental tests even if there is no causal connection. Statistical analyses must take such confounding factors into account.

Evidence that mental development is linked to breast-feeding eventually grew to the point where there was little room for doubt. By 1999, enough studies had been conducted to permit clinical nutritionist James Anderson and colleagues to carry out a sophisticated combined analysis of twenty previous studies. They took particular care to allow for confounding factors, such as socioeconomic status and maternal education level, to ensure that they homed in on the effect of breast-feeding itself. A significant advantage of breast-feeding emerged. When tested between the ages of six months and two years, breast-fed infants consistently showed higher levels of mental function than bottle-fed babies. And larger differences were found for premature babies than for infants born at normal weight. So the benefits of breast milk for mental development are even greater for preemies.

Another striking finding that emerged from the Anderson study was that benefits for mental development increased with duration of breast-feeding. Mother's milk is evidently better than formula for the infant's mental development, and breast-feeding for three years rather than a few months should yield even greater benefits. Yet "breast-fed" infants in comparative studies were often suckled for only a few months or even weeks. What we really need to do is to compare exclusively bottle-fed infants with babies breast-fed for a "natural" period of around three years. A valuable step in this direction was taken in a 1993 paper by developmental biologists Walter Rogan and Beth Gladen on breast-feeding and cognitive development. In a forward-looking study, they tested some eight hundred children at different ages between six months and five years. They found that scores were generally significantly higher in breast-fed children than in those that had been bottle-fed, although again only by a few points. More interesting was their finding that scores increased with the duration of breast-feeding, which ranged from a few weeks to a year or more.

Almost all evidence indicating that breast-feeding is advantageous for a baby's mental development is circumstantial. This is inevitable because ethical considerations generally rule out experiments of any kind. But one key experimental study does provide convincing evidence that supplementation of milk formula with the polyunsaturated fatty acids DHA and AA enhances mental development. In 2000, a team of researchers led by biologist Eileen Birch assessed the effects of adding DHA and AA for four months to a commercial milk formula fed to infants. This experimental approach eliminated many of the confounding factors that bedevil comparisons of breast-feeding with bottle-feeding. At four, twelve, and eighteen months of age, infants were assessed with standard developmental tests. For eighteen-month-old infants, adding both DHA and AA to formula resulted in an average increase of seven points on a standard scale of mental development. By contrast, no significant effects were found for muscle activity or general behavioral performance. This study convincingly establishes a causal connection between DHA and AA in human milk and brain development. The gun is no longer just smoking; it reeks of cordite.

While breast-fed infants have been shown to perform better on mental tests than bottle-fed babies, it has rarely been asked whether this advantage persists into adulthood. Filling this gap, in 2002 epidemiologist Erik Mortensen and colleagues published results from a long-term study of breast-feeding and IQ in more than 3,000 cases. In the study, the duration of breast-feeding was divided into five categories (less than one month, two to three months, four to six months, seven to nine months, and more than nine months), using information the mothers provided when their babies were one year old. Intelligence tests were conducted when those babies had become adults. Mortensen and colleagues took no fewer than thirteen potential confounding factors into account: social status and education of parents; marital status; mother's height, age, and weight gain during pregnancy; cigarette consumption during the last third of pregnancy; number of pregnancies; estimated gestational age; birth length and weight; and indicators of complications during pregnancy and birth. Even after allowing for all of these factors, the duration of breast-feeding was found to be significantly associated with increased adult scores in various intelligence tests.

BREAST-FED AND BOTTLE-FED babies differ in other important ways. Among other things, Bo Lönnerdal has reported that breast-fed infants have a lower weight at any given age than bottle-fed babies. Mothers often notice that, after a while, bottle-fed babies tend to be a little on the chubby side. One reason for this is a tendency to overfeed from a bottle, especially if the hole in the artificial teat is too large and allows the milk to flow rather than dribble. The perception of overweight babies corroborates evidence that bottle-feeding may be associated with a higher risk of obesity later in life.

Scientists have conducted studies to explore this possibility, but findings were initially inconsistent. Here, too, apparent conflict between results from different reports was satisfactorily resolved by a combined analysis. In a 2005 paper, obstetrician Thomas Harder and colleagues conducted an overall examination of seventeen studies that met a stringent set of conditions. They found strong support for a general decline in the risk of being overweight as the duration of breast-feeding increased. On average, every month of breast-feeding reduced the risk by 4 percent.

Breast-feeding also has other benefits for human newborns. One general effect is that breast-feeding provides comfort. Pediatrician Larry Gray and colleagues have shown that young infants respond less to pain when held and breast-fed by their mothers. In a routine hospital procedure, a blood sample is commonly collected with a lance from an infant's heel. In a randomized trial, fifteen newborn infants were held and breast-fed by their mothers during blood collection, while another fifteen babies were swaddled in their bassinets, in line with standard hospital practice. Crying, grimacing, and heart rate were all substantially reduced in infants that were breast-fed during blood collection. Thus breast-feeding evidently reduces the pain response in babies.

More dramatic is evidence that crib death is more likely in infants that are bottle-fed rather than breast-fed. First officially recognized in 1969 as sudden infant death syndrome (SIDS), the unexpected, symptomless death of an infant is a devastating experience for parents. In earlier times, when mothers often slept with their babies, crib deaths were generally attributed to "overlaying" and resulting suffocation. Even today, acute feelings of guilt are common when a crib death occurs, but parents are mostly blameless. It is now clear that SIDS is a genuine medical condition, although its causes remain obscure. What has been shown is that breast-feeding reduces the risk, although it remains to be seen whether the connection between milk composition and brain development plays a part in this. Other factors linked to bottle-feeding, such as the difference between rapid drinking from a bottle and slower milk intake from the breast, may be involved.

Environmental health researchers Aimin Chen and Walter Rogan examined the risk of crib death in relation to breast-feeding in the United States. They analyzed data from the 1988 National Maternal and Infant Health Survey, comparing more than 1,000 seemingly healthy babies who died between one and twelve months after birth with almost 8,000 infants who survived at least a year. Overall, the risk of dying during the first year of life was 20 percent lower if an infant had ever been breast-fed. Moreover, the risk declined as the duration of breast-feeding increased. The average risk for breast-fed infants was about 25 percent lower for infectious diseases, but only about 16 percent lower for SIDS—on the borderline of statistical significance. The authors concluded that promoting breast-feeding could potentially avoid more than seven hundred deaths during the first twelve months of life in the United States every year.

The link between bottle-feeding and the risk of SIDS was also examined by legal medicine expert Mechtild Vennemann and colleagues. This work drew on a German study of sudden infant death, comparing more than 300 infants who died of SIDS with 1,000 age-matched controls. Exclusive breast-feeding at one month of age was found to halve the risk of crib death throughout infancy. The team concluded by recommending that campaign messages aimed at reducing the risk of SIDS should include advice to breast-feed for at least six months.

Proteins differ between species, and the degree of difference is greater with distantly related species. Foreign proteins stimulate the body's defense mechanisms. So there is a possibility that allergic responses to proteins in artificial milk derived from a distantly related species such as the cow may play a part in crib death. There is, in any event, evidence that bottle-feeding may generally trigger allergic responses in susceptible individuals. Public health scientist Michael Burr and colleagues studied wheezing and allergy in almost five hundred children with a family history of allergic complaints. Wheezing occurred in just over half of children that had ever been breast-fed, whereas it affected three-quarters of exclusively bottle-fed children. The difference persisted even after allowing for several possible confounding factors. Burr and his colleagues concluded that breast-feeding may confer long-term protection against respiratory infection—yet another example of the benefits of suckling.

Breast-feeding an infant has real benefits for the well-being of the mother as well. Nursing immediately after birth helps reduce blood loss by increasing the tempo of uterine contractions. Starting from the knowledge that women experience afterpains when breast-feeding an infant, gynecologist Selina Chua and colleagues studied the influence of nursing and nipple stimulation on womb activity after birth. Their results showed that breast-feeding almost doubled the rate of womb contractions. Nipple stimulation alone also boosted contractions, although not to the same extent. Hemorrhage after birth is a major cause of maternal death in developing countries, so this finding has important practical consequences. More generally, breast-feeding ensures quicker recovery of the womb after birth and helps to restore the mother's general physical condition.

There are even more dramatic health benefits for the mother. Breast-feeding seems to have a protective effect, reducing the incidence of cancers of both the breast and the ovary. Records of mammals kept in zoos revealed that females that had never suckled offspring were more likely to develop mammary cancers. Paralleling this, investigations in the 1920s showed that human breasts that had never been used to feed an infant were more likely to become cancerous. In *Ever Since Adam and Eve*, reproductive biologists Malcolm Potts and Roger Short report that human breast cancer is about 120 times more common in industrialized societies than in gathering-hunting societies. In the United States, breast cancer currently occurs in one out of eleven women and kills one in sixteen. Statistical evidence also indicates that the risk of breast cancer declines as the number of infants suckled increases. The Collaborative Group on Hormonal Factors in Breast Cancer confirmed this in a large-scale, worldwide study published in 2002. The study included information from forty-seven epidemiological studies in thirty countries, covering a total of 50,000 women with breast cancer and twice that number without the disease.

However, there is a complicating factor in such studies. Pregnancy itself is known to provide some protection against breast cancer. Confirming this, the Collaborative Group survey showed that women with breast cancer had had 15 percent fewer births on average. The survey also showed that among women who had given birth, just over 70 percent of those who developed cancer had ever breast-fed. By contrast, almost 80 percent of women who remained cancer-free had breast-fed at some time, even if only for a few months. A further difference emerged for average lifetime duration of breast-feeding. On average, women who developed cancer breast-fed for only ten months in all, compared with fifteen and a half months for cancer-free women.

The most important finding was that the relative risk of breast cancer decreased by 7 percent for every birth and by more than 4 percent for every year of breast-feeding. The Collaborative Group combined all findings to estimate the cumulative incidence of breast cancer up to age seventy for developed countries. If all women had the average number of births and lifetime duration of breast-feeding that characterized third world countries until recently, deaths from breast cancer could be more than halved—from one in sixteen to one in thirty-seven. Breast-feeding accounts for almost two-thirds of this projected reduction.

Starting with studies carried out in the 1970s, breast-feeding has also been linked to a reduced risk of cancer of the ovary. A multinational study published in 1993 compared almost 400 cases of ovarian cancer with more than 2,500 carefully matched controls. In women who breast-fed for at least two months, the risk of ovarian cancer was reduced by about a quarter.

In this chapter, we have established that our infants are adapted for close contact with the mother and breast-feeding on demand. Still, it is important to remember that breast-feeding is but one part of mothering behavior, as will be shown in the next chapter. Close physical contact between mother and baby is itself a key part of mothering. Among other things, a breast-feeding mother must remain nearby to feed on demand. Mother-infant contact also depends on how the baby is carried from place to place. And the mother needs to care for her infant in ways other than breast-feeding, so there is a broader context to infant care to be explored.

CHAPTER 7

Baby Care: The Broader Picture

There is a lot more to human mothering than delivering milk. Baby care is a package deal including close, intimate contact and across-the-board attention to the infant's well-being. The basic primate pattern of suckling on demand depends on the mother's uninterrupted presence, and much of mothering behavior follows from that one simple requirement.

We must also remember that, as with other animals, the natural environment is very important. Like all other living things, humans were shaped by evolution under natural conditions, which must be considered as well if we are to fully understand ourselves. We must bear in mind the natural habitats in which human ancestors lived as they evolved. Settled communities are a recent development in human lifestyles, dating back only about 10,000 years. Before that, our ancestors lived as gatherer-hunters and were far more dependent on their natural environments. Gathering-hunting as a way of life prevailed for more than 99 percent of the time since our lineage separated around 8 million years ago from the sister lineage leading to chimpanzees.

* * *

In most nonhuman primates the mother carries the baby around as a passenger clinging to her fur. This act is a core feature of primate mothering. An infant is able to cling because it can grasp securely with its feet as well as with its hands. In the foot, the big toe closes against the other toes in a pincer-like action similar to the grip we use to hold something between our thumb and the other fingers. Because we humans became adapted to walk upright during hominid evolution, our feet gradually lost their grasping capacity, making us unique among primates. By contrast, the grasping power of our hand has been enhanced, leading to a peculiar reversal. Whereas we have a refined pincer-like action between the thumb and the rest of the hand, in other primates the grasping pincer is most obvious in the foot. Next time you visit a zoo, look at a young monkey or ape riding on its mother's fur and you will see how it grasps with its feet.

Generally, only the mother carries offspring. However, in some primates, including various New World monkeys, the father and occasionally other group members help carry infants. Squirrel-sized marmosets and tamarins typically have twins. Using clever experiments with added weights, behavioral biologists Gustl Anzenberger and Conrad Schradin showed with common marmosets from Brazil that the mother's energy costs will be significantly reduced when the father and other group members carry the twins. But in most monkeys and apes the mother carries a single infant almost constantly from birth until it starts to become independent. Even afterward, the infant occasionally returns to cling to the mother for suckling or security.

In all nonhuman primates the infant remains close to the mother while she sleeps, either clinging to her fur or snuggling against her in a nest. Continuously carrying a baby after birth guarantees a long period of intense mother-infant contact, providing special learning opportunities at the same time. All monkey and ape mothers are in direct contact with their infants virtually around the clock, so this was doubtless true of their common ancestor. Carrying infants had a long evolutionary history before humans evolved, so it is likely that our early ancestors followed suit.

Any parent knows that carrying babies demands effort and that the burden steadily increases as the infant grows. Energy costs for infant carriage are, after all, second only to the mother's energy burden for providing milk. For our highly mobile gathering-and-hunting ancestors, the demands of carrying infants surely would have been a significant part of any mother's energy budget. But it is no easy task to estimate precisely what the costs of infant carriage would have been. Fortunately, there is an indirect way to tackle this question, starting from a simple fact: As long as an infant is exclusively breast-fed, any energy it uses comes directly from the mother. Efficient use of resources therefore requires that a mother carry her infant whenever the additional energy expenditure is less than the infant would need to travel the same distance unaided. If the infant could cover the distance with a lower energy cost than the mother would face, riding on her would be wasteful.

This idea was the focus of a perceptive study of savannah baboons in Amboseli, Kenya, by field biologists Jeanne Altmann and Amy Samuels. These researchers carefully recorded how mothers carried their infants around. As expected, baboon mothers consistently carried young infants while traveling or foraging. For the first two months of life, mothers transported infants about a third of each day while covering a total distance of five to six miles. Thereafter, carrying gradually decreased and fell almost to zero by the time infants had reached an age of eight months. In the first months of life, rapid travel incurs especially high energy costs for the infant, so it makes most sense for a mother baboon to carry her offspring.

Anthropologist Patricia Kramer took a similar approach in a theoretical study in which she calculated energy consumption during walking for both mothers and children. Surprisingly, human energy use is particularly efficient during slow walking. At low speeds we humans are more economical than other mammals of similar body size. The costs of movement climb steeply at higher speeds, though, and other mammals are more efficient than we are when moving rapidly. Thus the energy advantage of human walking is limited to moving slowly, which is precisely what a mother does in most cases when carrying her baby. Just like baboons, human mothers use energy more efficiently than young infants when moving around. However, the extra energy cost of bearing a load depends on where it is borne on the mother's body. There is a smaller energy penalty if a load is carried near the body's center of gravity than anywhere else. (While it may be enjoyable, piggyback riding is not an optimal solution, biologically speaking.) Some evidence indicates that energy spent carrying loads may actually decrease with habituation.

As with baboons, the difference in energy costs between a human mother and her child rises as speed increases, so mothers should carry infants most often when moving rapidly. Interestingly, Kramer's calculations indicated that a mother should begin to force her child to walk most of the time at an age of about three years. On the other hand, at customary adult walking speed, no mother's energy budget would benefit if her child walked independently before the age of two. Mother-infant conflict is a potential problem here. As regards energy expenditure, it is always in the child's best interest to be carried. Even when the child walks independently, its optimal speed may differ from the most favorable speed for its mother. Thus mother and infant must reach some kind of compromise. Kramer concludes by noting that her findings may have significant implications for optimal intervals between births, suggesting a gap of four years based on her research.

As human evolution progressed, mothers eventually faced a second problem: The baby found it increasingly difficult to cling to its mother. Our hairy body covering, so convenient for grasping, was radically reduced somewhere along the line. An analysis of genes linked to hair growth indicates that human hair loss possibly occurred about 2 million years ago. Eventually, as a combined effect of losing both body hair and the grasping capacity of the foot, the evolving human infant ended up lacking both the ability to cling to its mother's body and the hair to cling to.

An alternative solution for carrying infants was therefore needed, leaving the mother's hands free for foraging and other activities. The first slings were made from naturally available materials such as lianas or animal skins. It has even been suggested that this development of baby-carrying slings led to the invention of clothing. An estimated origin for clothing has recently been set at almost 200,000 years ago. Genetic evidence indicates that body lice, which—unlike head lice—needed clothes to evolve, diverged from head lice at about that time. Whatever the case, even modern human populations with little or no clothing always have some contraption for carrying infants. The ancient practice of carrying infants surely remained part of our evolutionary heritage. The typical primate pattern of intensive mother-infant interaction fostered by close contact persisted as well. As we will see, though, there quickly arose another consequence of continuously carrying infants: the need to cope with infant waste.

IN AN INGENIOUS (if unappetizing) show of efficiency, most mammal mothers swallow the infant's waste products, including both urine and feces. Human mothers thankfully deal with the issue of waste in a different way. Gatherers-and-hunters have the easiest option, able to simply dispose of infant waste as they move around. However, once humans began living in settlements, that was no longer an option. There arose a need for an alternative disposal method, eventually resulting in diapers. Diapers have been used throughout recorded human history; several ancient Egyptian medical texts mention infants in diapers and even include advice on how to treat diaper rash. As the use of diapers spread, parents were faced with the vexing issue of toilet training.

Mammal mothers that give birth to poorly developed, altricial infants stop them from soiling the nest by swallowing their waste products. In mice and rats, for example, the mother often grooms her nestlings, licking them all over, and also swallows any urine and feces. Indeed, the mother's licking seems to prompt the offspring to discharge waste products. In fact, in some altricial mammals—such as certain carnivores—infants discharge urine or feces only when licked. This information is vital for successful hand-rearing. The mother's licking must be simulated by massaging the infant's lower belly with a damp cloth; without this, life-threatening retention of urine and feces can occur. At first sight, it is difficult to understand why carnivore babies are so reluctant to discharge their waste products. Perhaps carnivores are subject to particular pressure to avoid soiling the nest, and under natural conditions the mother ensures that the infants pass waste when she is around to deal with it.

Although they have well-developed, precocial infants, nonhuman primates typically show comparable toilet care. Mothers in the few nest-living species also keep their nests clean by swallowing waste products. This is true, for example, of nest-living mouse lemurs. In the course of my studies I reared several baby mouse lemurs, each smaller than my little finger, and used a warm damp cloth as described to provoke discharge of urine and feces.

In the primate species that carry infants continuously, the mother avoids soiling of her fur in a similar way. She swallows urine and feces, at least while the infant is quite young. Even our closest biological relatives, the great apes, do this. So this widespread mammal pattern must have disappeared at some stage during human evolution. Of course, we find the idea of swallowing infant urine and feces repugnant, but it's unclear how and when we abandoned this ancient mammalian behavior.

Because the mother ingests her offspring's waste, the fraught issue of toilet training simply does not arise with other primates. As they grow up, they gradually begin to discharge waste products while away from the mother. Like adults, they simply urinate and defecate wherever they happen to be. I experienced such carefree waste disposal firsthand while watching howler monkeys in the rainforest on Barro Colorado Island in Panama. From time to time, while I was sitting under a tree taking notes, a group of howlers above me would discharge, and urine and feces would rain down noisily through the leaves. I soon learned that little goes to waste in the complex network of a tropical rainforest. Within minutes, dung beetles buzzed up to exploit the rich harvest, rolling away the howler feces in small balls. An hour later, no trace of the deluge could be seen.

AT SOME STAGE in human evolution, we became picky about waste disposal and abandoned this freewheeling primate lifestyle. This eventually led to "potty training," a recent by-product of fixed human settlements. The outcome is that many mothers in industrialized countries now rely on using diapers in infancy and early childhood. Opinions differ wildly about the proper time to train an infant to use a toilet, veering to and fro between draconian timetables and extreme tolerance.

Ten successive editions of the U.S. government pamphlet *Infant Care* provide a quick historical overview of changing views regarding potty training. This all-time government best seller was first published by the Children's Bureau in 1914. Since then, more than 50 million copies have been sold. The 1935 edition specifically recommended an early start: "Training of the bowels may be begun as early as the end of the first month. It should always be begun by the third month and may be completed during the eighth month." As a training aid, mothers were advised to thrust a soap stick into the infant's bottom at precise times every day. An image of a clock alongside a mother and baby underlined the need for absurdly strict timing, with a permitted leeway of no more than five minutes. Incidentally, Luther Emmett Holt—that early champion of rigid feeding schedules for babies—similarly noted that potty training could begin in the second month after birth. In his 1894 book *The Care and Feeding of Children*, he also suggested using a piece of soap to help things along.

It gradually became widely accepted that highly regimented, coercive, or even punitive potty training can have severe emotional side effects. In 1942, for example, psychiatrist Mabel Huschka suggested a link between childhood neurosis and forcible bowel training. She specifically criticized the advice given in the 1935 edition of *Infant Care*. Freudians in general reported that such early, regimented training led to emotional difficulties and neuroses later in life. Eventually, authorities such as well-known pediatricians Benjamin Spock and T. Berry Brazelton swung the pendulum entirely the other way. They established the notion of a child-oriented approach to toilet training. As a result, the dominant child-rearing philosophy in the United States today is to let children decide when they are ready to use a toilet.

The current website of the American Academy of Family Physicians advises that a child is ready for toilet training when it can signal that its diaper is soiled or actually say that it needs to visit the potty. A child is usually ready for potty training at eighteen to twenty-four months, according to this view, but diapers may sometimes be needed up to the age of three. In the 1950s, all but 3 percent of children in the United States completed daytime potty training by the age of three. Nowadays only half of American children are potty trained by that age. Easy availability of labor-saving, disposable diapers is partly responsible for this shift. There is also a major downside for the environment—in the United States, millions of used diapers make up as much as a tenth of domestic waste.

Today, most American and European pediatricians believe that effective control of the bladder and bowel sphincters is simply impossible until the third year of life. However, comparisons across cultures reveal that ideas about readiness for potty training reflect social norms, not developmental constraints. A rarely cited 1977 paper by anthropologists Marten and Rachel deVries on the Bantu-speaking Digo of Kenya shows that the Digo began toilet training their infants in the first weeks of life. Without force or punishment, sensitive training ensured that around-the-clock bowel and bladder control was reached at an age of five to six months.

A Digo mother began by sitting on the ground with her legs held straight out in front. The infant was placed in a sitting position between her legs, facing away and supported by her body. She then made a special "shuus" sound that the infant learned to associate with urination. Success was rewarded by feeding or cuddling. The mother used a similar approach to encourage defecation but held her infant differently. The Digo example shows that early toilet training in infancy is entirely possible, given a suitably caring approach.

Digo toilet training in Kenya is by no means a unique example. There have been recent reports that mothers in China generally begin training their infants early as well, completing the process during the first year. As with the Digo, the mother gives an audible signal while holding the child over a latrine or open ground. Dutch writer Laurie Boucke cites these examples from Kenya and China, along with several others, in her book *Infant Potty Training*, first published in 2000 and already in its third edition. Boucke and Kathleen Chin coauthored a freely available Internet piece titled "Potty Training" stating that a human infant has a "window of learning" from birth to about six months of age. If parents utilize this window while it is still open, toilet training generally works well. As Chin and Boucke note: "The Western world has been indoctrinated to reject any form of early toilet learning. . . . Millions of happy babies in China can't be wrong!"

A YOUNG HUMAN BABY cannot cling to the mother's body, and the welfare package the mother provides no longer includes swallowing waste. However, the mother's deep-rooted primate heritage of carrying infants still persists. Close physical contact between mother and baby remains crucial for healthy development. It not only ensures warmth and security but also is profoundly important for mother-infant bonding. Enlightened maternity units now tacitly recognize this. A newborn infant is handed to the mother to cuddle as soon as possible after birth rather than being kept in isolation.

Human babies, like infants of other primate species, are undoubtedly preprogrammed to expect intensive physical contact. Absence of such contact is a major shock to the system. Placing a human baby in a separate crib, let alone in a separate room, for much of the day and all of the night is a radical departure from a standard primate pattern with an 80-million-year history.

WHAT HAPPENS IF THINGS GO WRONG? If a mammal mother fails to suckle properly, infant survival and therefore reproductive success are threatened. You might think that natural selection would favor any adaptation that promotes successful infant rearing. It seems only logical that, once a mother has invested in pregnancy, she should protect and foster her investment after birth. But things are not quite so simple. Under certain conditions, mammal mothers may cut their losses. In some extreme cases, the mother eats her offspring soon after birth. This practice is common in mammals such as rodents and treeshrews, which have altricial offspring that are small and poorly developed at birth. Young children are often horrified and traumatized when a cuddly pet hamster gives birth to a litter of pups and then promptly devours them. Yet this is the mother's natural response to adverse conditions.

A mother's daily investment of resources is even higher when suckling than during pregnancy. Therefore, not rearing infants under adverse conditions may well be a cost-cutting strategy; it may be preferable to give up on a litter and save resources for offspring born under better conditions. From this perspective, cannibalism of infants can be seen as a gruesome kind of recycling. This behavior is very rare or absent among mammals that give birth to precocial offspring, including primates. Mothers may abandon or even kill precocial infants, but they almost never eat them. Perhaps maternal investment during long pregnancies in precocial mammals is simply too heavy for cannibalism to be a viable cost-saving strategy. If so, then mothers of precocial infants should be expected to show adaptability that will allow infants to survive even under adverse conditions. This is indeed the case.

Suckling is especially sensitive to adverse conditions. Many studies have shown that it is vulnerable to stress triggered by negative factors in the physical or social environment. Treeshrews provide a graphic illustration. They are monogamous in the wild and thrive in captivity only when kept in pairs. If two or more adults of the same sex are kept in the same cage, social stress gives rise to problems. Behavioral physiologist Dietrich von Holst conducted a study of this very effect. It so happens that treeshrews are well suited for such research: The extent to which they fluff their tails reflects increasing levels of stress hormones circulating in the blood. Thus tail fluffing can be used to judge a treeshrew's stress level from afar. Using this hands-off approach, von Holst showed that in treeshrews suckling is the first element of female reproduction that is disrupted by stress. At higher stress levels, cannibalism, pregnancy failure, and eventually complete shutdown of the ovary occur.

Occasional stress can also reduce suckling in treeshrews without suppressing it altogether, as I learned from an incidental observation when I was studying a breeding colony of treeshrews at University College London. Treeshrew infants are ideally suited for studies of milk intake because the mother visits them only once every two days. So I set about measuring the amounts of milk swallowed by treeshrew nestlings by weighing them before and after each suckling visit by the mother. Everything went according to plan until, while monitoring one particular litter, I noticed something odd. The regular two-day suckling rhythm was disrupted on two occasions a week apart. The mother actually visited the nest more often at those times, but overall daily milk intake was reduced and the nestlings ended up seriously underweight.

Concerned, I asked the manager of the animal facility whether he had noticed anything out of the ordinary. His facility was mainly used to breed mice and rats, so I asked whether anything unusual had happened with those animals as well. His eyes lit up as he provided the explanation. He had noticed in the past that breeding of the mice and rats was disrupted by fire alarm tests—a full minute of white noise from a large bell. Because of this, he had a standing arrangement that he would be warned of any upcoming test, allowing him to remove the metal cover from the alarm in good time. Yet a new and overzealous safety officer had tested the alarm without the usual warning. For good measure, he tested the alarm once more a week later. The timing of the two tests exactly matched the suckling disruptions I had noticed with my treeshrews. The facility manager then showed me breeding records for the mice and rats during the same period. Sure enough, there was a sharp fall in numbers of offspring recorded at the time of each fire alarm test. Likely the usual numbers of offspring were born, but many died or were devoured. Clearly, an abrupt stress such as an unexpected loud noise can disrupt breeding even in long-domesticated laboratory rodents.

Primates, with their well-developed precocial infants, respond similarly to stress, albeit with a few important distinctions. In captivity, primate mothers often fail to rear their infants. Milk shortage is at least partly to blame. Infants simply waste away unless removed in time and reared by hand.

Biologist Christopher Pryce, while working as my PhD student in London, investigated the maternal behavior of Amazonian red-bellied tamarins, small New World monkeys that typically have twins. Following the usual pattern for primate colonies, some mothers reared their infants while others did not. Over the course of his research, Pryce identified a hormonal difference between successful and unsuccessful mothers. The startling thing was that the difference was already detectable in midpregnancy. In other words, it was possible to predict fairly accurately from hormone levels halfway through pregnancy whether a mother would go on to rear her infants or not. This finding was one of the first clear indications that maternal failure in primates might have some hormonal basis.

Scientists need to know whether hormones influence mothering in primates, because of possible implications for humans. It is often assumed that mothering in primates—or at least in monkeys, apes, and humans—depends heavily if not exclusively on learning. Some zoos, for instance, have installed in the cages of chimpanzees and gorillas videos of mothers suckling infants, optimistically hoping that the viewing primates might learn to rear their babies properly. Such actions reflect the widespread notion that a mother's failure to rear her infant is generally due to social influences. When mothering fails in primates, stress is not often considered as a contributory factor.

There is a simple explanation for the widespread belief that primate mothering depends mainly on learning. It stems from the well-established fact that maternal behavior often improves with successive births, particularly after the first. As a general rule among primates, successful mothering is more frequent with females that have given birth previously (multiparous mothers) than with those giving birth for the first time (primiparous mothers). The improvement is generally attributed to learning, but there could be other explanations. For instance, the first birth is likely to be especially stressful for any female, and underlying physiological mechanisms may need tuning. From an evolutionary standpoint it seems hardly likely that mothering, which is fundamental for reproductive success, would depend exclusively on learning and have no physiological safeguards.

STARTING IN 1957, psychologist Harry Harlow at the University of Wisconsin–Madison carried out a series of controversial experiments on rhesus monkeys. The results of these studies, which are superbly reviewed by Deborah Blum in *Love at Goon Park*, dramatically influenced interpretations of primate mothering. The research team removed newborn monkeys from their mothers to study effects of maternal deprivation. Harlow designed rudimentary "surrogate mothers" to which an infant could cling. In one famous experiment, he isolated a newborn infant from its mother in a separate cage and gave it a choice between two surrogates, a bare wire frame or a similar frame covered with soft cloth. A bottle attached randomly to one of these contraptions provided milk. Infants clung mostly to the cloth-covered frame, regardless of whether it had a milk bottle. If the bottle was attached to the bare wire frame instead, an infant moved to it only briefly to drink. When frightened, it always retreated to the cloth-covered frame, even when the milk bottle was on the bare frame.

Harlow concluded that cloth-covered frames provided some comfort for isolated infants, but certainly not enough for normal development. Infants that had access only to a simple wire frame had problems digesting milk and had diarrhea more often. When later housed with other monkeys, individuals raised with surrogates turned out to be misfits. They were withdrawn and antisocial, repeatedly banging their heads and rocking. As a result, such motherless monkeys couldn't mate normally when they became sexually mature. So Harlow intervened. Using what he callously called a "rape rack," he restrained adult females and subjected them to forced mating. If pregnancy and birth ensued, these motherless mothers neglected or even abused their infants. Harlow's interpretation of this behavior was that lack of maternal bonding during their own development prevented these females from engaging properly with infants. They failed to form any attachment and infant rearing was disrupted. Yet in some cases mothering did improve after experience with a series of infants. Harlow's experiments and conclusions have been widely criticized. But it has been equally widely accepted that a female infant must be properly mothered for her to become a successful mother in her turn.

Primatologist Maribeth Champoux revisited the topic of motherless mothers in rhesus monkeys in a 1992 paper, noting that there might be another explanation for the problems that Harlow reported. Flawed mothering could be influenced by adverse effects of social isolation during infanthood, she argued, not just lack of mothering. Mothering styles vary widely even among normally reared female macaques. Furthermore, motherless monkeys can become adequate caretakers after experience with infants. Champoux compared normal, mother-reared infants with some that had been separated but raised in groups (peer-reared) and others that had been completely isolated (isolate-reared). Peer-reared mothers did have less contact with their infants than mother-reared females, but isolate-reared mothers showed more actual rejection of infants. The extreme effects of complete isolation are somewhat mitigated by living alongside other motherless infants after separation from the mother.

In the 1980s I happened to meet wildlife enthusiast Peggy O'Neill when she participated in a summer school that I co-organized at the Durrell Wildlife Conservation Trust. She had inherited rhesus monkeys surviving from Harlow's experiments and had managed to transform them into a thriving free-ranging troop in a large outdoor enclosure. Under these changed conditions, the motherless mothers overcame many of their previous social defects. As it turned out, they were able to show quite competent mothering. I realized that it was possible, under the right conditions, to cancel out at least some of the deficits caused by the stress of early isolation.

IN MY OWN EXPERIENCE, observations over the course of years have convinced me that failure of mothering in monkeys and apes might be due more to presence of stress than absence of learning. In the 1980s, surveys of great apes in captivity—chimpanzees, gorillas, and orangutans—revealed that only half of infants born were successfully reared by their mothers. For that reason, many babies were removed for hand-rearing. This is still a major problem for breeding great apes in zoos and other institutions. The traditional explanation has been that great apes in zoos were often captured or isolated as youngsters and reared under unsuitable social conditions. As a result, it is said, they never learned the skills needed for successful mothering. In this scenario, separation and hand-rearing of great ape infants create a vicious circle, passing poor mothering from one generation to the next.

However, some hand-reared great apes have reared their own infants impeccably. It is clearly not essential to grow up in a social group to be a successful mother. I began to suspect that stress from living in cramped, unsuitably furnished cages might play a part in maternal failure. This suspicion was reinforced when I witnessed an interesting development during my long-term studies of gorilla reproduction at Durrell Wildlife Conservation Trust. Two breeding females kept together in the same cage with a single adult male had each experienced three pregnancies. After every birth, the infant was removed within a few days because mothering was not working out. The gorilla group was then moved from its rather cramped quarters to a much larger, state-of-the-art facility with an extensive, well-furnished outside enclosure. Soon afterward, both breeding females gave birth again. From the outset, the mothers reared their infants properly. Neither female had had any opportunity to learn adequate mothering before giving birth in the new facility, so what had happened? One strong possibility is that the original cramped cage was stressful and that the spacious new quarters relieved that stress and any accompanying hormonal effects.

For quite some time afterward, I cherished the idea of launching research on gorillas to study hormonal changes during late pregnancy and the weeks after birth. My aim was to find out whether there are detectable hormonal differences between successful and unsuccessful mothers. Eventually, my PhD student Nina Bahr at the University of Zürich conducted a project that met this goal. Nina painstakingly collected samples of urine and feces from nine captive female gorillas for several weeks before and after birth in each case. She then looked for connections between hormonal patterns and infant rearing. Nina found that the quality of mothering was, indeed, associated with patterns of the steroid hormones estrogen and progesterone and of the stress indicator cortisol. Considerable further work is needed to test for an actual causal relationship between stress, hormones, and mothering in great apes. But this hypothesis seems at least as likely as the traditional notion that social learning mainly or exclusively drives successful mothering.

Some research, such as that reported in a 1997 paper by psychologist Alison Fleming and colleagues, has revealed similar associations between hormones and mother-infant interactions in women. Fleming's team had two aims: to use questionnaires to find out whether human mothers undergo a change in maternal responsiveness in late pregnancy, just before birth, as reported for other mammals, and to determine whether there was any relationship between hormonal change and changes in maternal feelings and attitudes. In this part of the study, Fleming also conducted assays of various steroid hormones. Nurturant feelings increased during pregnancy and became even stronger after birth. The study also found that attachment feelings were associated with proportional change in estradiol and progesterone from early to late pregnancy. Fleming noted that any association between hormones and attachment may be explained in two ways: by a direct influence on nurturant feelings or by indirect effects on a general perception of well-being. Here, too, more work is needed to show whether a direct causal connection exists.

* * *

Women are often unpleasantly surprised when they feel an emotional letdown after childbirth. Thankfully, this is usually no more than mild and short-lived "baby blues," which affect more than half of new mothers. However, some women experience persistent postpartum depression severe enough for them to need medical attention. Symptoms usually start soon after birth, but the onset can occur months later. On top of tiredness and insomnia, these symptoms include inexplicable sadness, crying bouts, social withdrawal, loss of appetite, anxiety, and irritability. Postpartum depression is a devastating and far too common condition; although it was first recognized as a medical complaint in the 1850s, it has been regularly screened only for the past twenty years. In the industrialized world, it affects one in seven new mothers. Making matters even worse, affected mothers often feel guilt and shame, sometimes to the point of feeling suicidal. Needless shame often holds them back from seeking the professional help they so urgently need.

"Once upon a time, there was a little girl who dreamed of being a mommy. She wanted, more than anything, to have a child. . . . [O]ne day, finally, she became pregnant. She was thrilled beyond belief. She had a wonderful pregnancy and a perfect baby girl. At long last, her dream of being a mommy had come true. But instead of being relieved and happy, all she could do was cry." This passage is how Brooke Shields opens her frank, personal account of postpartum depression in her moving 2005 book *Down Came the Rain*. Courageously, she went public to warn others about a condition that can afflict any woman.

There is a long list of risk factors linked to postpartum depression by different studies. These include poverty, low social support, prior history of depression, poor marital history or single parenthood, unplanned pregnancy, low self-esteem, complications of labor, use of anesthesia during birth, hormonal imbalance, stressful life events, and cigarette smoking. The list also includes bottle-feeding instead of breast-feeding.

In a 2006 paper, gynecologist Sarah Breese McCoy and colleagues compared 81 women diagnosed with postpartum depression with 128 women identified as not depressed. Of the risk factors considered, the greatest effect was seen with bottle-feeding. This behavior was associated with a significantly higher risk, more than double that for breast-feeding mothers. Women with a prior history of depression also showed a significantly higher risk, increased by almost 90 percent, while cigarette smoking was associated with a risk almost 60 percent greater. Although the risk of postpartum depression was somewhat higher in women who gave birth through Caesarean section, the difference was not statistically significant. Effects of some factors identified were found to be additive.

A link between bottle-feeding, hormones, and postpartum depression had previously been identified by a medical team in the United Arab Emirates led by Mohammed Abou-Saleh. This research group studied seventy women a few weeks after birth, examining levels of various hormones in blood samples. Estrogen, prolactin, and cortisol were all higher in women who had just given birth than in nonpregnant controls. However, women diagnosed with postpartum depression had significantly lower prolactin levels than new mothers without depression. Women who developed depression six to ten weeks after birth also had significantly elevated progesterone levels. Mothers who bottle-fed had lower prolactin levels than women who breast-fed their infants, and the former were significantly more likely to have postpartum depression. Abou-Saleh and his team also found that women with a prior history of depression had significantly lower prolactin levels and were more likely to suffer from postpartum depression.

It is now widely accepted that postpartum depression is triggered by major hormonal changes during late pregnancy and following birth. Evidence indicates that maternal feelings and effectiveness may be linked to a balance among hormones, particularly estrogen, progesterone, and prolactin. These hormones have been used with some success to treat postpartum depression. The good news is that treatment with a combination of medication and therapy is often extremely effective. It takes a while, but women do usually recover.

Everything that I have learned from studying nonhuman primates has led me to the conviction that stress plays a part in triggering postpartum depression in human mothers. Stress disrupts hormone balances, and that, in turn, can affect the physiology of mothering. If this is indeed part of the story, we really need to work at lowering the stress exposure of new mothers. There are clear indications that postnatal depression is less likely if a mother breast-feeds, but exposure to stress may adversely affect the production of milk. Thus problems must be tackled on a broad front.

* * *

STRESS IS NOT the only factor to be considered. There is an even more striking example of a vital but often neglected natural mechanism involving breast-feeding. A neat feedback arrangement links nursing frequency to fertility. Folk wisdom stretching back at least as far as Aristotle tells us that a woman is less likely to conceive while nursing an infant. This advice suggests that suckling suppresses ovulation. Yet so many women have become pregnant while breast-feeding that this notion has often been dismissed out of hand. Scientific detective work, notably by reproductive biologists Peter Howie and Alan McNeilly, eventually set things straight. It turns out that suckling does effectively inhibit ovulation in women, but only if the infant is breast-fed regularly around the clock. The basic pattern of suckling on demand is a twenty-four-hour occupation. If a human baby is not suckled during the night, inhibition of ovulation is greatly weakened. So around-the-clock suckling is needed for reliable, long-term suppression of ovulation after birth.

Suppressing ovulation by suckling provides a simple way to link fertility to the mother's nutritional status. If the mother herself is well nourished, her milk will be relatively rich. The infant she suckles will be quickly satisfied and probably will wait some while before seeking another session on the breast. By contrast, milk produced by a poorly nourished mother will be more dilute. The infant is hence likely to suck for longer in any one bout and to become hungry again relatively quickly. Infants of poorly nourished mothers will be suckled more often than infants of well-nourished mothers. Therefore suppression of ovulation after birth will last longer in an undernourished mother, as long as she suckles around the clock. This is a truly elegant natural feedback mechanism.

Experiments with red deer reported by Andrew Loudon and colleagues in 1983 have also shown this kind of mechanism in action. Infants of mothers kept on rich pasture are suckled less often and the mothers become fertile again more rapidly compared to mothers kept on poor pasture. Suckling of human infants is doubtless adapted in a similar way. Provided that suckling occurs by night as well as by day, it has a natural contraceptive effect that can last for a year or more after birth. Indeed, British biologist Roger Short asserts in a 1976 paper that worldwide more births are prevented by breast-feeding than by all other forms of contraception combined.

Having broached the topic of natural contraception, it is now time to look at the various ways in which we intervene in our own reproduction today. Armed with a broad understanding of the biological background to human reproduction, we can more confidently interpret the special issues arising from artificial manipulation.

CHAPTER 8

Monkeying with Human Reproduction

In 1798, Thomas Malthus—a British theologian and mathematician—published *An Essay on the Principles of Population as It Affects the Future Improvement of Society*. This essay triggered the key notion of natural selection for both Charles Darwin and Alfred Russel Wallace and paved the way for their theory of evolution. Malthus noted that if there are no constraints, human populations expand geometrically over time. Geometric increase is like compound interest; numbers expand at an accelerating rate. By contrast, the best we can manage with food supply is a basic annual increase, like simple interest. Any human population growing unchecked will double in size every twenty-five years and sooner or later outstrip its food supply. Severe competition is the inevitable outcome. Poverty, famine, and, in many cases, armed conflict follow. Darwin and Wallace realized that a struggle for existence arises because all animal and plant populations have a natural tendency to expand geometrically and outstrip available resources. In a competitive environment, natural selection favors inherited features that improve the chances of survival.

Malthus opposed both contraception and abortion. He proposed instead that late marriage combined with sexual restraint could limit human population growth. Yet, despite his own admirable example—marrying at thirty-eight and having only three children—the world continues to suffer from the fundamental problem he so astutely spotted. Without planned intervention, human populations will continue to grow like compound interest, inexorably outstripping available resources. Moreover, unbridled human population growth has already led to catastrophic levels of environmental pollution. The current world population, after quadrupling over the past century, is now around 7 billion. United Nations figures have projected continued growth by another 30 percent, to reach over 9 billion by the year 2050.

Thanks to Malthus and countless successors over the past two centuries, the menace posed by the population explosion is now widely recognized. It is morally unacceptable to stand by as the global human head count continues to climb, at a staggering cost in human hardship and environmental destruction and pollution. Yet, with rare exceptions, governments have been reluctant to intervene in an area where religious beliefs, economic factors, and political expediency are all major obstacles. Ultimately, the population crisis poses an important question about what "natural" means: Is it natural to accept that human populations will grow geometrically until they reach the point where famine, poverty, and armed conflict overwhelm us all? Or is it natural to apply our outstandingly large brains to the problem of controlling population size to reduce the global burden?

DELIBERATE INTERFERENCE in our own breeding is unique to humans, with a history stretching back thousands of years. Unlike the previous topics covered, there is no evolutionary precursor to birth control. Intervening in human procreation is decidedly *not* natural. All the same, as in many other instances, it is vital to understand the natural background. Successful intervention in our own breeding demands a sound grasp of underlying biological processes, particularly when there may be negative side effects. Furthermore, this is a realm in which religion and government often intervene, imposing moral or legal boundaries and regulating procedures. Ethical arguments about human reproduction are often based on what is supposedly natural. But how far do suppositions match up with biological reality? That is one of the key issues I tackle in this chapter.

Intervention in human reproduction cuts two ways. In one direction, biological know-how can be used to limit breeding with various methods of birth control. In the diametrically opposed direction, that same know-how can be used to combat infertility. What an ironic reflection of the human condition—while countless couples are anxiously seeking ways of avoiding pregnancy, others are desperately trying to have a child.

Medical advances have opened up a wide array of possibilities for both birth control and assisted reproduction. However, many people reject some or all of the available procedures, often for religious reasons. Let's be clear: Freedom of religious belief and practice is a basic human right that must be vigorously defended. At the same time, dilemmas can arise when belief and biological reality collide.

In some cases religious objections to birth control and assisted reproduction have gradually softened, but in others they have stubbornly persisted. For instance, the Church of England's Lambeth Conference in 1920 still condemned all "unnatural means of conception avoidance." Just ten years later that same body approved the use of birth control by married couples. The U.S. Federal Council of Churches followed suit eleven years later. But the circular letter *Casti Connubii,* a response issued by the Roman Catholic Church in 1930, implacably opposed "artificial" contraceptives of any kind. That position remains unchanged to the present day.

This chapter will consider birth control first and then turn to assisted reproduction. While both topics raise fundamental biological and moral issues, that sequence reflects my desire to end the chapter on an upbeat note, emphasizing interventions that can bring great happiness to couples struggling to achieve parenthood.

The simplest birth control methods, which came first historically, stop sperms meeting eggs. The most basic method of all is abstinence. Total abstinence—lifelong celibacy—is the strictest and safest form of contraception, reducing an individual's genetic contribution to the next generation to zero. In contrast to total celibacy, the less drastic option of partial abstinence may be seen as a relatively harmless, natural way of limiting births. As Malthus suggested, copulation might be delayed as long as possible and then restricted to times when conception is desired. This practice does not work in the real world. Alternatively, conception may be avoided without abstinence by withdrawal before ejaculation (coitus interruptus), though this unsettling practice has a notoriously high failure rate. One in four women relying on this method will conceive in the course of a year.

A common form of contraception that requires neither abstinence nor withdrawal is some kind of physical barrier that catches the sperms before they meet the egg. Popular methods include a condom used by the man or a cap, diaphragm, or sponge inserted into a woman's vagina. Condoms, the modern human counterpart of Spallanzani's taffeta pants for frogs, have been used for at least four centuries. They began with sheaths made of animal membranes, especially lamb intestine or bladder. The earliest reliable written record refers to linen sheaths pretreated with a chemical solution, made in the mid-sixteenth century by Italian anatomist Gabriele Falloppio. (The oviduct is called the "Fallopian tube" in his honor.) Over the course of the nineteenth century, use of male condoms gradually increased to become the most popular method of birth control in the world. The first rubber condom was produced in 1855 and latex versions were introduced in 1920, although "lambskin" condoms made from intestines are also available.

Simple barrier methods all have a double problem, however: They disrupt intimacy and have a high failure rate in practice.

What do failure rates signify? Two different annual rates are now usually reported. *Perfect use* refers to completely correct and consistent application of a particular method. *Typical use* is a catchall measure including incorrect and/or inconsistent application. Perfect use is pie in the sky; typical use is contraception as used by real people. Population researcher James Trussell and colleagues reported in 1990 that in typical use the failure rate of condoms is 12 percent. Diaphragms, caps, and sponges have a failure rate around 18 percent. Perfect use reduces the failure rate to 3 percent for condoms and 6 percent for diaphragms. With all barrier methods, success can be somewhat improved through combined use of sperm-killing agents (spermicides). When used alone, spermicides have an especially high failure rate, up to 30 percent, but they can boost the effectiveness of standard barrier methods.

Condoms have an incidental benefit that has recently increased radically in importance: reduced transmission of sexually communicable diseases. In the past, the greatest scourge of this kind was undoubtedly syphilis, which mercifully has become less virulent over time. The earliest documented strain first appeared in a European outbreak in the 1490s. Symptoms were severe and often fatal within a few months. Falloppio, who published a treatise on syphilis in 1564, reported test results showing that his linen sheath protected men against the disease. Extensive experience since that time has shown that condoms at least reduce transmission of sexually communicable diseases. AIDS is the current sexually transmitted scourge, and condoms are proving effective in reducing infection.

There exists another special form of barrier contraception for women that is placed in the womb—the intrauterine device (IUD). Unlike use of other barrier methods, inserting an IUD is a delicate procedure that needs medical help, and regular checkups are mandatory. Any foreign object in the womb provokes an inflammatory response, creating a hostile environment for sperms, so IUDs probably act mainly by preventing fertilization. One early IUD was a flexible silk ring supported by a gold wire, designed by gynecologist Ernst Gräfenberg in 1928. Later, more effective versions of the Gräfenberg ring were made with silver wire. But greater effectiveness was actually due to a copper impurity, almost one part in four. Modern IUDs are popularly known as "coils," although many are T-shaped. They are usually plastic and often coated with copper. IUDs are now the world's most widely used reversible method of birth control. Around 160 million women currently use them. Two-thirds of users are in China, including half of all married women. Typical-use failure of modern IUDs is quite low, around 1 to 2 percent per year. However, in rare cases where pregnancy does occur, the IUD must be removed as quickly as possible because its presence increases the risks of miscarriage or premature birth.

Barrier methods of contraception are undeniably unnatural and often provoke objections. In some eyes, use of a condom, withdrawal to prevent fertilization, and masturbation are all "sinful." Such wastage of sperms is quaintly described as "spilling the seed," evoking mental images of bungling gardeners. Male masturbation, in particular, has often been seen as corrupt behavior that must be suppressed at all costs. Yet it is quite common among nonhuman primates, especially monkeys and apes. It occurs casually and attracts no obvious response from other group members. In *Primate Sexuality*, Alan Dixson lists no fewer than thirty-five species of monkeys and apes that have been seen to masturbate, twenty of them under natural conditions. It is not true, as sometimes claimed, that masturbation is an aberration of monkeys and apes kept in captivity.

The fact of the matter is that men continually shed vast numbers of surplus sperms in their urine. Moreover, prolonged abstinence generally leads to spontaneous emission during sleep. Thus there is nothing inherently unnatural about nonreproductive "seed spilling." Indeed, abstinence has the same end effect as masturbation. A thought-provoking 1975 essay by physiologist Roy Levin suggested that both masturbation and nocturnal emissions serve important functions by maintaining the volume of stored seminal fluid within the normal range and reducing the occurrence of abnormal sperms. Levin noted that many mammals other than primates, both wild and captive, engage in masturbatory activity or show spontaneous discharges. Examples include shrews, hamsters, rats, cats, dogs, deer, bulls, horses, and whales. A recent study in Australia indicated that regular ejaculation in men is important for maintaining sperm quality.

Moral principles regarding wastage of sperms and eggs are closely tied to concerns about the sanctity of life. It is perhaps understandable that sperms are seen as living entities in the complex process that eventually leads to birth of a human baby. Western thought on this matter has been influenced by a long history in which a single sperm has been regarded as more or less equivalent to a human individual. Three centuries ago, when microscopy was in its own infancy, some investigators claimed that the sperm head contains a miniature human being, a homunculus. In 1694, Dutch mathematician and physicist Nicolaas Hartsoeker published an influential drawing of a homunculus bunched up in a sperm head. As he freely admitted, the drawing showed not what he saw but what he thought should be there. The emerging "spermist school" held that after fertilization of an essentially passive egg, development occurs as the homunculus carried in the head of the sperm unfolds. This interpretation owed much to ingrained prejudices identifying males as active and females as passive, in reproduction just as in society at large.

Yet male chauvinists of the time did not have it all their own way. An alternative school of thought, ovism, favored a diametrically opposed interpretation. Ovists claimed that the egg contains the active principle in development, while the sperm plays a minor part as a catalyst at fertilization. Supporters of both schools of thought—collectively called preformationists—believed that miniature human beings, residing in either sperms or eggs, were products of original creation. However, they blithely ignored a fundamental and inescapable problem in their reasoning: infinite regress. If a sperm or egg contains a preformed miniature human being, that entity must presumably enclose even tinier sperms or eggs containing yet smaller preformed human beings, and so on ad infinitum. In her 1997 book *The Ovary of Eve*, biologist Clara Pinto-Correia aptly likened preformation to a set of Russian dolls. Although an egg has far more room than a sperm head, even ovists had a problem with staggeringly large numbers of preformed individuals. In 1766, Albrecht von Haller, taking the age of the Earth as 6,000 years, calculated that Eve's ovary must have contained 200 billion preformed entities. Preformationism also faces a classic problem with hybrids between species: If sperms or eggs of horses and asses contain miniature, preformed individuals, how can matings between them produce mules?

Quite apart from hybridization and the Russian-doll paradox, it is difficult to see any merit in spermism. The average human ejaculate contains around a quarter of a billion sperms, and every male typically produces many trillions of sperms during his lifetime to father a comparatively tiny number of offspring. Of the 250 million sperms in the average human ejaculate, only a few hundred actually end up at the fertilization site in the oviduct. Given these chances, it would be unbelievably wasteful for any sperm to carry anything more than the bare essentials needed to contribute to the next generation. For sperms, fertilization is a gigantic lottery. In the vast majority of cases, regardless of context, the success rate is zero. In the few instances where copulation actually leads to conception, one out of a quarter of a billion sperms succeeds.

In his 2006 book *Sperm Wars*, reproductive biologist Robin Baker reported that about 500 copulations take place for every child born. In other words, for every successful conception around 125 billion sperms are ejaculated during copulations. Over the lifetime of any individual human male, stupendous numbers of sperms are inevitably consigned to oblivion one way or another. Spermism demands the tragic loss of trillions of homunculi for every man that ever lived. Yet this concept lives on today in the curious notion that wastage of human sperms violates the sanctity of life. Wastage happens all the time. It is a fact of life; it is natural. Of course, detailed research with progressively improved microscopes eventually revealed that sperms and eggs do not contain preformed miniature humans patiently awaiting their cue for development. Instead, each sperm and each egg contains half of the DNA that is needed at fertilization to restore the full genetic complement required for development. In that sense, then, eggs and sperms are incomplete entities. Once released, they have no independent existence beyond a few days.

One widely used method of avoiding conception has widely escaped religious opposition: abstinence when ovulation is likely. The explicit aim is to identify a woman's "fertile window" and then restrict copulation to the remaining "safe period." Such planned abstinence is deliberately designed to waste sperms and eggs, but it requires no special devices or treatments, so it is commonly accepted as a relatively natural kind of birth control. Indeed, it is the sole method recognized as morally acceptable for Roman Catholics. Pope Pius XII gave it an official seal of approval in 1951, and it is prevalent in certain regions of the world.

Planned abstinence builds on the standard "egg timer" model of the human menstrual cycle discussed in Chapters 1 and 3. It envisages a clock-like monthly cycle, with ovulation occurring about halfway between one menstruation and the next. But until the 1920s it was widely believed that the most fertile time in the menstrual cycle coincided with menstruation, not with midcycle, and women were actually encouraged to restrict copulation to midcycle to avoid conception. The notion of midcycle ovulation was first built into ovulation-tracking calendars in the 1920s by Japanese obstetrician Kyusaku Ogino and Austrian surgeon-gynecologist Hermann Knaus. Birth control based on their ideas later became known as the Ogino-Knaus or Knaus-Ogino method, ironically contracting to either OK or KO. In 1930, physician John Smulders—a Roman Catholic—explicitly recommended abstinence at midcycle as a way of dodging conception. This recommendation reflected a general belief that sperms can survive only for about two days, while egg survival is limited to a day at most. Smulders's advocacy led directly to Catholic approval of the method. As H. L. Mencken wryly noted: "It is now quite lawful for a Catholic woman to avoid pregnancy by a resort to mathematics, though she is still forbidden to resort to physics or chemistry."

Calendar-based birth control is popularly known as the rhythm method. The label probably owes its origin to Leo Latz, a Chicago physician whose classic book *The Rhythm of Fertility and Sterility in Women* was published in

1932 and reprinted twenty-six times. The foreword explains how Latz introduced the method in the United States, with the stated aim of having a system that any health professional could teach in three minutes. Latz developed basic rules to identify fertile and infertile phases of the menstrual cycle. The simplest version instructs a woman to regard cycle days twelve to nineteen as the core fertile window. For greater security, she is then told to calculate the difference between the longest and shortest of her last eight to twelve cycles and to add that number of days to the front end of the window. For example, if a woman has cycles lasting between twenty-six and thirty-one days over the course of a year, she should tack on five days, extending the fertile window to cover days seven through nineteen. If you had difficulty following this calculation, you are not alone.

The rhythm method may be seen as natural, but it has a basic flaw. In typical use, it is unreliable. Over the course of a year roughly one in five women using the rhythm method will become pregnant. A 1996 review of well-designed clinical studies by reproductive biologists Robert Kambic and Virginia Lamprecht estimated an average failure rate of about 15 percent per year. In fact, in typical use the failure rate is sometimes as high as 25 percent, scarcely better than for withdrawal. It is not surprising, then, that this method has become pejoratively known as "Vatican roulette."

The rhythm method has been refined in various ways to try to improve the success rate, notably by physician John Billings. In a 1981 paper, he noted that secretion of mucus from the neck of the womb reliably indicates ovulation time. Thus a woman can potentially identify fertile and infertile phases in each menstrual cycle by checking wetness and slipperiness of the vulva and any visible discharge of mucus. Billings, a staunch Roman Catholic like Knaus and Latz before him, was reportedly guided by his religious faith in developing this method of family planning, and in 1969 his efforts were rewarded with a papal knighthood. The Billings Ovulation Method rapidly attracted a considerable following, especially in Australia, and the current official website claims a failure rate of only 1.5 percent. An authoritative 2011 survey by population investigator James Trussell, however, reported a failure rate of 3 percent for the ovulation method in perfect use and an average of 24 percent for all planned abstinence methods in practical use.

Attempts have been made to refine the periodic abstinence method, but it remains a stubbornly unreliable birth control technique. The simplest, one-size-fits-all prescription is abstinence on specific days of the cycle. As we have seen, though, average cycle length differs from one woman to another, and so Latz recommended that each individual carefully track her menstrual cycles to identify her own basic rhythm. This reflects the notion that there is a range from slow to rapid cycling and that every woman has her own egg timer setting. Individualized tracking of this setting is known as the calendar method.

As shown in Chapter 1, many studies have shown that average cycle length varies markedly from woman to woman. Because of this wide variation, an individualized calendar method should increase the chances of successfully identifying the fertile window. This is made more complicated because cycle length is not constant for any particular woman, varying greatly from one cycle to another.

A USEFUL ADJUNCT to the calendar method identified about a century ago is measurement of basal body temperature (BBT), the lowest value for temperature while resting. BBT is commonly increased by between 0.5°F and 1°F after ovulation, remaining at that level for the rest of the cycle.

Unfortunately, the BBT rise typically occurs a day or two after ovulation. It cannot be used to predict ovulation time in an ongoing cycle; it only permits wisdom after the event. Instead, temperature records collected over several cycles may be used as an additional aid to estimate the typical ovulation time for any particular woman. But detection of ovulation with BBT is imprecise, with an uncertainty of plus or minus three days, and does not always work. Gynecologist Kamran Moghissi examined relationships between BBT and hormones across one menstrual cycle in thirty normally menstruating women, concluding that one in five women showed no BBT rise despite showing a normal hormonal pattern.

Even with faithful adherence to methods of planned abstinence, many conceptions still occur during the designated safe period. One reason is that the standard egg timer model of the human menstrual cycle is an average picture; it is a statistical abstraction, not biological reality. Moreover, the rhythm method relies entirely on avoiding the most likely time of ovulation. It is assumed that copulation must occur close to ovulation to result in conception. As Chapter 3 showed, although ovulation generally occurs close to midcycle, this is not true of fertile copulation. Conception can result from a single copulation on virtually any cycle day. The likelihood of conception is highest during the first half of the cycle (follicular phase) and comparatively low during the second half (luteal phase). In fact, information from single copulations indicates that they are more likely to lead to conception around day ten than on midcycle days fourteen or fifteen.

A MORE URGENT ISSUE is that of timeworn sex cells. According to the egg timer model, both ovulation and conception typically occur at midcycle, whereas copulation can occur at any time. Sperms and eggs have limited life spans, so only copulation close to ovulation should result in conception. If the egg timer model is correct, a recurring danger of fertilization with timeworn sex cells is inevitable. If copulation occurs more than two days before ovulation, a decaying sperm might fertilize a fresh egg just released from the ovary. If copulation occurs more than a day after ovulation, a freshly ejaculated sperm might fertilize a decaying egg that is on its way to the womb. Experiments with laboratory mammals have shown that fertilization with decaying eggs or sperms can result either in miscarriage or in fetal abnormality if pregnancy proceeds. Yet this potential problem has received surprisingly little attention from scientists when it comes to ourselves, even though it is accepted that humans copulate throughout the cycle.

It is possible that humans—along with monkeys and apes, which also commonly copulate over an extended period in the cycle—have a special adaptation that somehow reduces or eliminates risks of fertilization with decaying sperms or eggs. Perhaps there is a filtering process that identifies stale sex cells and stops them in their tracks. Alternatively, there may be a mechanism that blocks development of any embryo resulting from fertilization with a timeworn egg or sperm. However, these possibilities have not been investigated. We simply do not know whether fertilization with timeworn sex cells occurs in humans. Nor do we know what the outcome might be. As things stand, we are left with a neglected problem with serious implications, particularly for the rhythm method.

If the rhythm method works as planned and conception is avoided, all is well and good. But if the method fails and unplanned conception occurs, a delay between copulation and ovulation is virtually guaranteed. Copulation limited to a supposed safe period must surely increase the likelihood that any conception that does occur will involve a timeworn sex cell. Fertilization will occur either when a sperm survives just long enough to meet up with a freshly released egg or when an egg survives just long enough to encounter a freshly ejaculated sperm. In either case, Vatican roulette appears in a more sinister light.

Fluctuating cycles render the rhythm method unreliable for birth control. Although available evidence indicates that ovulation close to midcycle is fairly typical, cycle length and timing vary greatly. Yet the standard way of illustrating hormone patterns creates an illusion of regularity. Precisely because cycles are known to vary greatly in length, hormone levels are not plotted by cycle day counting from menstruation. Instead, they are centered around ovulation time, usually detected from a spike in the hormone LH. Ovulation time is defined as cycle day 0 and other days are numbered plus or minus on either side. Visually, this suggests a degree of precision for ovulation time that is lacking in real life.

Gynecologist Maria Elena Alliende specifically set out to compare individual hormone profiles of twenty-five women of prime reproductive age with the average picture for the menstrual cycle. Three or more cycles were monitored by BBT and mucus, and morning urine samples were collected daily for hormone measurements. The overall hormone profile closely resembled the picture reported from previous studies, with an average cycle length of twenty-eight days. Yet in three-quarters of the women, individual patterns differed substantially from the average profile. Clear peaks in estrogen and LH were not always present. Some cycles showed continuous high levels, while others had multiple peaks, making ovulation time uncertain. Only progesterone profiles were relatively regular.

But there is an even bigger problem. If sperms, eggs, or both can survive longer than is generally accepted, copulation leading to conception need not coincide with ovulation. Survival of an egg beyond a day or so seems highly unlikely. As discussed in Chapter 3, single copulations leading to conception occur frequently throughout the follicular phase but are relatively uncommon during the luteal phase. Thus extended sperm survival is likely to be largely or exclusively responsible for any difference in timing between ovulation and copulation leading to conception. If sperms survive beyond the generally accepted two-day period, they must be stored somewhere in the womb or oviducts. Blind, branching crypts in the neck of the womb not only produce mucus but can also provide temporary refuge for hundreds of thousands of sperms. While in the crypts, sperms remain partly or completely immobile. They are then gradually released over a period of at least five days and possibly longer. Yet sperm storage in the crypts has been little studied, leaving many questions unanswered. For instance, it remains unknown whether sperms released from crypts after more than a few days are still fertile. The additional possibility of short-term storage of sperms in the oviduct also deserves proper exploration.

As regards the rhythm method of birth control, the difference between the standard egg timer model and an alternative model with extended sperm survival is important in one key respect. If the standard model is correct, then midcycle abstinence should increase the risk of a timeworn sperm fertilizing a freshly ovulated egg. By contrast, extended sperm survival would reduce or eliminate this risk. Either way, problems arising from timeworn eggs would be expected to arise. Midcycle abstinence should increase the risk of conceptions arising from fertilization of decaying eggs by either fresh or stored sperms.

And now the really crucial point: The rhythm method of birth control may be expected to increase the incidence of conceptions involving decaying eggs and, perhaps, timeworn sperms. The prediction from the standard egg timer model actually gives more cause for concern. If sperms can survive for some time in a woman's reproductive tract, the outlook regarding frequency of conceptions involving timeworn sex cells is less worrisome. In that case, any increase should be limited to decaying eggs that survive just long enough to be fertilized by insemination timed to avoid the fertile window. The logical conclusion from all this is the prediction that midcycle abstinence will in any case carry a higher risk of developmental accidents. If the standard egg timer model of human midcycle ovulation and conception is correct, the risk should be substantial. Even if sperms survive for an extended period in a woman's reproductive tract, the risk, although smaller, will still be present. A saving grace here is that any sperms stored from copulation before ovulation are probably more likely to fertilize the egg when it is released than are any sperms from copulation after ovulation.

* * *

One predictable outcome of fertilization with timeworn sex cells should be increased numbers of miscarriages. It is difficult to get any solid numbers to test this expectation, however, as early miscarriage, occurring within the first four weeks after conception, can easily pass unnoticed. Minor bleeding can easily be misinterpreted as delayed or irregular menstruation. As studies of laboratory mammals have shown, abnormal chromosomes are common in timeworn sex cells. In 1970, obstetricians Rodrigo Guerrero and Claude Lanctot reported a possible link between decaying sex cells and miscarriage in women. It has been estimated that almost three-quarters of human conceptions fail at some stage. Most are lost in early pregnancy and only one in ten ends in a clinically recognizable miscarriage.

Animal studies also suggest a more serious possibility: Midcycle abstinence might increase the risk of developmental abnormality in human pregnancies that continue to term. This phenomenon has been little studied, but papers published in the 1960s and 1970s by gynecologist Leslie Iffy provide valuable information. Spurred by an interest in abnormalities of human pregnancy, notably development of the fetus at an inappropriate site (ectopic pregnancy), Iffy explored possible links to likely conception time. Direct information on conception dates was lacking, so Iffy used standard tables to assess developmental age of newborn babies, then reckoned backward to estimate conception time. In 1963, he reported that estimated conception dates for ectopic pregnancies clustered mainly after midcycle, in the luteal phase. However, his results contained an even bigger surprise. Several estimated conceptions fell into the luteal phase *before* the last recorded menstruation. In other words, conception seemingly occurred in the cycle before the one in which pregnancy usually would be assumed to have started. Comparable results were found with other fetal abnormalities.

To account for these results, Iffy proposed what he called a "reflux theory": Menstruation occurring after conception might displace the embryo before implantation, flushing it into the belly cavity or displacing it downward in the womb. A subsequent paper published by Iffy and Martin Wingate in 1970 explicitly addressed the issue of potential risks associated with the rhythm method of birth control. They found that there was a significant statistical association between use of this method and increased incidence of various kinds of fetal abnormality.

Another key paper by James German noted a potential link between delayed fertilization and the incidence of Down syndrome. It is generally accepted that the incidence of Down syndrome, caused by a well-defined chromosomal abnormality, tends to increase in mothers over the age of thirty-five. This effect is usually attributed to declining egg quality in older women. German made the interesting alternative proposal that, with advancing age, declining copulation frequency might play a part. Analysis of hospital data confirmed the predicted pattern. German did not discuss any connection with the rhythm method of birth control, but the implication is clear: Any increase in fetal abnormality that may be due to a decline in copulation frequency could also arise with deliberate abstinence at midcycle.

EPIDEMIOLOGIST PIET JONGBLOET has explored possible links between timed abstinence and developmental defects. He examined various indirect associations between contraceptive failures and births of infants with abnormalities, particularly chromosomal aberrations, noting that the incidence of Down syndrome was more than doubled in young Catholic mothers. The highest frequencies of Down syndrome were found in the Irish Republic and in Roman Catholic populations in the Netherlands and Australia. Moreover, an investigation published in 1978 and 1981 by geneticists Iva Milstein-Moscati and Willy Beçak had shown that conceptions of children with Down syndrome are generally preceded by unusually long periods of abstinence. A retrospective study comparing mentally retarded offspring with siblings also revealed that abnormality was increased when parents using the rhythm method copulated exclusively during the period after midcycle. Jongbloet concluded that his findings support the hypothesis "that post-midcycle conceptions and other contraceptive failures during application of the so-called 'natural' methods of family planning harm the resulting progeny."

Such potential problems are not confined to Roman Catholics. Reproductive biologist Teresa Sharav published a paper examining timeworn sex cells in relation to the incidence of Down syndrome in Orthodox Jews living in Jerusalem. Orthodox Jewish women follow very strict rules regarding reproduction: They practice abstinence from the onset of menstruation until they have undergone cleansing in a ritual bath (*mikve*) seven days after the last trace of blood. With a regular twenty-eight-day cycle, this would amount to stopping abstinence on about day twelve, quite close to ovulation. Yet fertilization may occur belatedly if menstruation is overlong, if the cycle has a short follicular phase, or if a woman delays her visit to the *mikve*. This ritual offers a unique opportunity to study births in relation to conception time. Sharav found that the incidence of Down syndrome among women younger than thirty-seven years of age was significantly higher among Orthodox Jewish women than in the nonobservant population. She concluded that fertilization following delayed copulation could explain these results. In any event, her findings indicate another case in which prescriptions regarding the timing of abstinence in the menstrual cycle may be associated with a higher incidence of birth defects.

That conclusion fits with a 1979 report coauthored by reproductive biologist Ernest Hook and medical scientist Susan Harlap. They showed that the incidence of Down syndrome was almost doubled for Jewish mothers of Asian and North African origin compared to the average level elsewhere. These incidences rank among the highest anywhere in the world. By contrast, Jewish women of European origin had Down syndrome rates no higher than those observed in the general population in the United States or northern Europe. This can be explained by diminished observance of traditional practices by Jewish women in Europe.

All of these findings sparked initial flurries of interest, but there has been little follow-up research. As things stand, there is enough circumstantial evidence to give real cause for concern. Taken together, published findings lead to one important conclusion: In cases of unintentional conception, deliberate abstinence around the time of ovulation may lead to a higher incidence of miscarriage and a higher frequency of fetal abnormalities.

STARTING IN 1986, obstetrician Joe Leigh Simpson set out to assess safety of the calendar method. His large-scale collaborative study involved six centers with extensive experience of natural family planning. For a large sample of women, the likely day of ovulation was estimated from cervical mucus or raised basal body temperature. Conception time was inferred from individual charts in which each woman recorded days with copulation. It was simply assumed that conception always resulted from copulation closest to the likely date of ovulation, thus potentially biasing the results. Conceptions were divided into those associated with copulation at an "optimal" time and those linked to "nonoptimal" copulations. Only the previous day and the actual day of ovulation were regarded as optimal; all other cycle days were treated as nonoptimal.

In a 1997 overview, the Simpson consortium reported that rates of clinical miscarriage and fetal abnormality were not significantly higher with nonoptimal conceptions. However, one significant difference was reported: With nonoptimal conception, miscarriage was about three times more common in women with a previous history of pregnancy loss. Nevertheless, the authors drew the following overall conclusion: "Our findings should be reassuring to natural family planning users." This conclusion was widely and positively reported in the media, feeding a belief that timeworn sex cells are not a problem with the calendar method of birth control. Yet the conclusion is unjustified because it is not appropriate to restrict "optimal" copulation to just two days around the likely time of ovulation. With calendar methods, copulation is avoided for between ten and fourteen days, not just for two days. Concerns about timeworn sex cells apply to the front and back ends of a long abstinence period, not to the immediate vicinity of ovulation. The Simpson consortium was asking the wrong question.

In 2006, philosophy professor Luc Bovens published a thoughtful essay on possible implications of the rhythm method for embryonic development. He wrote his essay in light of the fact that pro-life advocates oppose any birth control method that may result in embryonic death. His analysis began with the reasonable assumption that the viability of an embryo will decrease as the interval between insemination and ovulation increases. Bovens correctly noted that conception resulting from insemination at the tail ends of the fertile period is especially likely to be associated with embryonic defects. The predicted outcome is that the rhythm method should lead to increased occurrence of spontaneous abortions resulting from filtering of embryonic defects. Bovens concluded that "the rhythm method may well be responsible for a much higher number of embryonic deaths than some other contraceptive techniques."

A MAJOR REVOLUTION in birth control for women began just over fifty years ago. It was sparked by development of hormone preparations that could be taken conveniently by mouth. The oral contraceptive pill—widely known as the birth control pill or simply the pill—delivers precisely gauged daily doses of steroid hormones. Most versions combine progesterone- and estrogen-like substances. It had long been known that large doses of certain steroids block ovulation in mammals. Those hormones disrupt natural feedback control of the ovary by blocking release of the hormones FSH and LH from the pituitary gland. The main effect is that development of follicles ceases and ovulation is suppressed. The strategy to tap into this natural mechanism to control fertility is really quite ingenious. Throughout human pregnancy, circulating hormones—particularly progesterone—block ovulation. Simulation of hormone levels found in early pregnancy can therefore stop a nonpregnant woman from ovulating. In other words, the birth control pill misleads a woman's body by imitating early pregnancy. It is a prime example of exploiting a natural mechanism to beneficial ends.

In fact, there exists a plant counterpart to the birth control pill. While studying the droppings of two troops in a free-living population of baboons in Nigeria, biologist James Higham and colleagues detected marked seasonal peaks in progesterone-like by-products. Detective work revealed that both troops consumed a particular food only at times when those by-products peaked—fruits and young leaves of the African black plum. Laboratory tests confirmed that this plant contains high levels of progesterone-like substances. Indeed, peaks of by-products in the droppings of female baboons were much higher during periods of feeding on African black plum than peaks of actual progesterone derivatives during pregnancy. The unusually high levels of progesterone-like compounds in the diet suppressed sexual swelling, an external indicator of an active ovary. Moreover, no conceptions occurred during peak feeding on African black plum. Consumption of this plant seemingly has a contraceptive effect, simulating pregnancy like some forms of the birth control pill.

Humans have also made use of such plant-based contraceptives. A close relative of the African black plum found in the Mediterranean region of Europe has been linked to increased levels of progesterone-like compounds in the blood. Common names for the Mediterranean species include "chasteberry" and "monk's pepper," reflecting its use in medieval times to throttle sexual desire in men of the cloth. But for more than two millennia, dating back to ancient Egypt and Greece, chasteberry has mainly been used to treat gynecological disorders. One prominent application has been to stimulate menstruation, and modern clinical studies have identified definite benefits for treating menstrual disorders and infertility. Medical historian

John Riddle cited evidence that chasteberry has long been used for contraception as well. Dioscorides—the leading first-century authority on ancient pharmaceuticals—stated in his treatise *De materia medica* (*Materials of Medicine*) that chasteberry "destroys generation as well as provokes menstruation." It seems that a limited amount of chasteberry stimulates menstruation, whereas high doses block conception. Here we have an early herbal equivalent of the pill that, in small doses, helped monks keep their vows of celibacy. What could be more natural than that?

Medieval monks may have drawn inspiration from Pedro Julião, a Portuguese physician confusingly known as "Peter of Spain." In 1272, he published a hugely popular book entitled *Thesaurus pauperum* (*Treasure of the Poor*). Starting with concoctions aimed at dampening sexual desire, the book went on to counsel the poor on birth control. Julião listed various herbal preparations, including chasteberry, to be taken by mouth. In 1276, this early advocate of herbal birth control became Pope, taking the name John XXI. Sad to say, the one and only medical Pope occupied this position for only eight months. He insisted on continuing his scientific studies even after his election, and had a special annex built for this purpose. One day, the ceiling collapsed and killed him.

SO THE PILL HAD herbal forerunners. Yet two obstacles held back practical development of the modern contraceptive pill. Most important, large doses are needed if steroids are taken by mouth. Extraction of steroid hormones from animal tissue is far too expensive, so cheap sources had to be found. Russell Marker, an organic chemist at Pennsylvania State University, solved this problem when he discovered that large quantities of progesterone can be synthesized from plant extracts. His breakthrough came when he found that this extraction could be done relatively cheaply from an inedible yam growing in rain forests around Veracruz, Mexico. Subsequently, several chemists, including Carl Djerassi, designed ways of producing various synthetic progesterone-like compounds in the laboratory. The second obstacle to developing an oral contraceptive was sheer apathy laced with reluctance. Government agencies, university research departments, and the pharmaceutical industry all dragged their feet. The National Institutes of Health (NIH), for instance, banned funding for birth control research until 1959.

Without the help of NIH, reproductive physiologist Gregory Pincus initiated serious research in 1951, encouraged by Margaret Sanger, founder of the American birth control movement. It soon proved possible to proceed to real-life tests. Start-up funding from philanthropist Katharine Dexter McCormick enabled fruitful collaboration with John Rock, a Roman Catholic gynecology professor at Harvard University, and his colleague Min-Cheuh Chang. In fact, Rock's interest at the outset was not in birth control but in treating women for infertility. This collaboration emphasizes once again that reproductive know-how can be used to boost as well as block fertility. Rock's approach was to give oral doses of steroid hormones to women for several months to establish and maintain a pregnancy-like state. He would then stop the treatment. Thanks to a bounce-back effect (the "Rock rebound"), one in six previously infertile women became pregnant. Rock tested the first version of Pincus's contraceptive pill in the same way, and similar numbers of rebound pregnancies occurred when he stopped treatment.

Although Rock's main concern was treatment of infertility, some Roman Catholics bitterly attacked him. Monsignor Francis Carney of Cleveland called Rock a "moral rapist," while another American obstetrician actually sought out Cardinal Richard Cushing in Boston to have Rock excommunicated. Rock taught clinical obstetrics for more than thirty years at Harvard Medical School in Massachusetts, a state in which distribution of contraceptives was at first legally banned. In 1963, he published his book *The Time Has Come: A Catholic Doctor's Proposals to End the Battle over Birth Control*, which was widely discussed and often vilified.

The first field trials of oral contraception in women were conducted in Puerto Rico in 1956. The version of the pill used contained steroids synthesized by Djerassi and other chemists: a progesterone-like compound combined with an estrogen. The Federal Drug Administration granted initial approval for use of oral contraceptives in the United States in 1960. Thereafter, the pill rapidly became a popular form of birth control around the world. Currently, oral contraceptives are used by some 12 million women in the United States and by more than 100 million women worldwide. In Great Britain, a quarter of women between the ages of sixteen and forty-nine now use the pill. Surveys have shown that birth rates are halved in populations that use the pill.

Daily doses of hormones in any version of the pill must be carefully gauged to achieve the desired results. In most cases, the dosage sequence is designed to simulate a twenty-eight-day cycle. This usually includes a week without the progesterone-like compound to ensure that regular menstrual bleeding occurs, although this is not medically essential. As anthropologist Beverly Strassmann noted, many gynecologists unfortunately believe that women have to menstruate every month.

There are two basic pill sequences. With a twenty-one-pill package, a pill is taken every day for three weeks and then no pills are taken for a week. The alternative twenty-eight-pill package similarly has a sequence of twenty-one hormone-containing pills, but seven sugar pills containing no hormones are tacked on the end to maintain daily pill taking. In either case, protection from conception continues during the hormone-free week. A more recent development is a three-month version of the pill with a constant daily hormone dosage. Menstruation occurs less frequently with this version.

The argument that the birth control pill works by simulating a natural ovulation-blocking mechanism has been much debated. All oral contraceptives containing a progesterone-like compound have one secondary effect: They inhibit sperm penetration through the neck of the womb by decreasing the amount of cervical mucus and making it more viscous. Thus oral contraceptives might be seen as agents that also hinder sperm migration. Moreover, it has been suggested that the pill can induce changes in the inner lining of the womb that block implantation of an embryo. For this reason, some have claimed that the pill might provoke early abortion. However, oral contraceptives are specifically designed to block ovulation with steroid hormones. The likelihood of fertilization during their use is very low to nonexistent. Therefore any incidental barrier to sperm migration is irrelevant, and very early abortion is highly unlikely. Nevertheless, in his influential 1968 encyclical *Humanae Vitae* (subtitled *On the Regulation of Birth*), Pope Paul VI interpreted oral contraceptives as immoral. To John Rock's bitter disappointment, the pill was lumped with other methods of birth control as "artificial." In 2008, Pope Benedict XVI reaffirmed this official Roman Catholic stand.

One undoubted advantage of the pill as a method of birth control is that it is extremely reliable. With typical use of the pill, annual pregnancy rates range between 2 and 8 percent, while the rate with perfect use is less than a third of 1 percent. The method is hence unquestionably more reliable than barrier contraception or any kind of planned abstinence. Benefits arising from the reliability of the pill are not limited to more effective control of family size. One striking finding is that legalization of the pill in any given country is soon followed by a marked increase in the rates at which women participate in higher education and study through to graduation.

THERE HAVE BEEN WIDESPREAD CONCERNS that pill use may be associated with health risks. There were several reports in the late 1960s that blood clots, strokes, and heart attacks are more likely. As a result, in the United States the number of women using the pill was halved in the decade from 1975 to 1984. It does seem likely that oral contraceptives can adversely affect blood coagulation. This effect may predispose to clotting in the lungs and deep veins, increasing the risks of stroke and heart attack. Because of this risk, the pill is generally not prescribed for women with preexisting cardiovascular disease, an inherited predisposition to form blood clots, severe obesity, and/or high cholesterol. Its use is also regarded as unsuitable for women smokers over the age of thirty-five. A potential effect of the pill on body weight has also been an issue. Here, the jury is still out: Some studies indicate that its use leads to mild weight gain; others suggest that mild weight loss occurs.

Regardless of any reported findings, one point is absolutely clear: Any health risks associated with the pill are much lower than the usual risks accompanying the pregnancies and births that occur without birth control. Moreover, pill use has several health benefits. One notable, if minor, advantage is that it reduces the incidence of acne. Indeed, the pill is sometimes used to treat this condition in young women, in which case contraception is an incidental side effect. More seriously, birth control pills are known to alleviate various medical conditions including menstrual irregularity, premenstrual syndrome, and pelvic inflammatory disease. Even more important, use of the pill for five years halves the likelihood of cancer of the ovaries or womb later in life, and risk reduction continues as pill use extends beyond five years.

Possible links between hormone-based contraception and the risk of breast cancer have also been much discussed. Research historically revealed complex interrelationships and conflicting results. Yet all became clear in 1996 with results from a large-scale reanalysis of data from the Collaborative Group on Hormonal Factors in Breast Cancer for more than 150,000 women from fifty-four individual studies. It was established that there is indeed a small increase in the likelihood that breast cancer will be diagnosed while women are taking combined oral contraceptives. However, cancers diagnosed in women who are using, or have used, combined oral contraceptives are clinically less advanced than those diagnosed in women who have never used the pill. It is likely that women who seek gynecological advice for birth control also benefit from increased screening and hence early detection of breast cancers.

Pill usage has also been challenged from a different direction. Much has been written about possible environmental effects of steroids from millions of pills ending up in sewage. Both natural and synthetic steroids, particularly estrogens, are excreted in the urine and feces. It has been suggested that they may cause hormonal disruption in wild fish populations living in water contaminated by treated sewage effluents. Several studies have suggested that estrogens discharged into the environment feminize male fish and lower their fertility. Unsurprisingly, doctrinaire opponents of the pill have identified environmental pollution as an additional manifestation of its evil workings. It must at once be noted that orally administered estrogens and progesterone-like compounds are not used only for contraception, of course. They are also used to treat hormonal disorders and for replacement therapy in postmenopausal women. Moreover, steroid hormones are widely used in animal farming, notably as growth promoters. It should also be remarked that in late pregnancy women naturally produce massive quantities of estrogens and substantial amounts of progesterone, which are continually voided in urine and feces. This point is hardly ever mentioned in all of the fuss about environmental pollution and the pill.

As is so often the case, detailed research has revealed a complex mesh of interactions. A large-scale collaborative study by the Universities of Brunel, Exeter, and Reading in England confirmed that polluted water disrupts hormonal systems in fish. But it emerged that nonsteroidal chemical agents released into rivers and lakes after sewage treatment inhibit the function of testosterone in male fish. Such antiandrogens probably play a key part in feminization. Other research has indicated that exposure to anti-androgens can also damage human reproductive health. It now seems that a complex cocktail of interacting chemicals, rather than estrogens and progesterone-like compounds alone, is responsible for hormonal disruption.

* * *

With the exception of condoms, development of contraceptive methods for men has generally been entirely eclipsed by methods designed for women. As noted in Chapter 1, one simple approach to male contraception, reputedly having a long tradition in the Middle East and particularly in Turkey, is to heat the testes. Under normal conditions, sperms are produced and stored at a temperature below the core body level. Raising the temperature of the scrotum, even for short periods of time, can impair fertility for several weeks. Scrotal temperature can be raised by applying hot water, generating heat with ultrasound, or wearing special undergarments that hold the testes close to the abdomen. Various experiments, some conducted by John Rock, indicate that heating testes is safe, effective, and reversible. However, we do not know whether long-term use has negative health impacts or continues to affect sperm quality after cessation. Given the simplicity of this approach, it is little short of amazing that so little research has been conducted. This topic surely deserves serious attention.

Possibilities for developing a male pill have been much discussed, but a practically usable method remains an elusive prospect. In a 1994 commentary in *Nature,* Carl Djerassi and reproductive physiologist Stanley Leibo assessed prospects for developing a male pill beyond 2010 as dismal. Development, testing, and regulatory approval of a new oral contraceptive takes fifteen to twenty years, so nothing is likely to change for many years to come. In short, there is every justification for the feminist complaint that development of contraception focuses primarily on monkeying with women.

Attempts have been made to right this imbalance. For instance, a protocol involving injections of synthetic progesterone to block sperm production was developed, combined with application of a testosterone gel to offset side effects. Unfortunately, this treatment reduces sex drive and has other unwelcome downsides such as weight increase and fatigue. Other research has aimed at interfering with sperm maturation in the epididymis. The chemical compound phenoxybenzamine was found to block ejaculation without affecting semen quality. Moreover, the effects are reversed when treatment is stopped. But this method has not been developed further, doubtless because it blocks ejaculation altogether. In 1929, a study conducted in Jiangxi, China, revealed that cooking with crude cottonseed oil was associated with low male fertility. The contraceptive effect was traced to the chemical compound gossypol. Pills made from gossypol, which is found in okra as well as cottonseeds, inhibit the action of certain enzymes. The Chinese government conducted trials of this agent for some fifteen years. Although gossypol has a reliable contraceptive effect, it also seriously affects health and causes permanent sterility in 10 to 20 percent of users, so the effort to develop a male contraceptive was eventually abandoned in 1986.

In recent years, developmental biologist Michael O'Rand has explored another approach to reversible male contraception. He and his team conducted a series of experiments with rhesus monkeys, injecting them with antibodies against a protein called eppin that is specifically located on the surface of human sperms. Male monkeys immunized in this way are rendered infertile. Moreover, tests showed that anti-eppin antibodies significantly decreased the motility of human sperms. This research raises the prospect that injection of appropriate antibodies could eventually be used as a male contraceptive for humans.

Methods of contraception discussed thus far are generally reversible. As a rule, no significant reduction in fertility occurs once a person stops using them. Yet anyone who has decided to forgo any prospect of future reproduction can resort to sterilization, which is virtually irreversible. In men, the sperm-carrying duct (vas deferens) from each testis can be severed in a procedure known as vasectomy. The equivalent for women is tying off the oviducts (tubal ligation).

For men, vasectomy is a minor surgical procedure designed to exclude sperms from the ejaculate. The testes remain in the scrotum, where they continue to produce sperms along with testosterone and other male hormones. Sperm-laden fluid from the testes makes up less than 10 percent of a normal ejaculate, so vasectomy reduces ejaculate volume only a little, and everything else remains largely unaffected. However, sex drive is reduced in at least one in ten vasectomized men, and roughly the same number regret deciding to have the operation in the first place. In any case, before undergoing vasectomy it is advisable to have sperm samples collected and preserved. As a contraceptive method, vasectomy has very low failure rates. Within a few months after vasectomy, the rate of unintended pregnancy falls to less than 1 percent. Worldwide, about 6 percent of married couples practicing birth control now rely on this method.

For women, tubal ligation is a form of permanent sterilization in which both oviducts are severed, cauterized, or clamped shut. The aim is to block fertilization of any egg released from the ovary. As with vasectomy, surgery is required, although several methods employ a relatively limited intervention using a laparoscope. Hormone production, menstrual cycles, and sexual desire can all be affected by this procedure. As with vasectomy, in the first year after operation the failure rate of tubal ligation is only 1 percent. Effectiveness may decline slightly as the years pass because oviducts sometimes re-form or reconnect. Unfortunately, if this does occur, there is a one in three chance that implantation will occur in the wrong place (ectopic pregnancy). Worldwide, among married couples relying on sterilization, tubal ligation is about five times as common as vasectomy.

Surgical removal of both ovaries, known as ovariectomy, sometimes along with the oviducts, is a more drastic method of female sterilization that is rarely used for birth control. Removing the ovaries leads to major hormonal changes and accompanying side effects resembling intensified symptoms of menopause. Women are generally advised to use hormone replacement therapy after the operation. Alarmingly, cardiovascular disease becomes seven times more likely. The risk of premature loss of bone density (osteoporosis) also increases.

On a global scale, surgical sterilization now counts among the most widely used birth control methods. In China, almost 40 percent of married couples have one sterilized partner. Elsewhere, proportions are lower. In the United States, one in three couples rely on female sterilization for contraception, so it is the commonest birth control method.

THE MOST EXTREME form of permanent sterilization is undeniably castration. This term is often reserved for males, reflecting the relative simplicity of cutting off descended testes compared with the challenge of surgically removing the ovaries from the belly cavity. In medical science, though, a distinction is made between male castration (orchidectomy) and female castration (ovariectomy).

Accounts of castration as an act of devotion are included among the earliest archaeological records of human religion. Finds from the Neolithic site of Çatalhöyük in southern Anatolia (Turkey) indicate that ritual self-castration for goddess worship occurred almost 10,000 years ago. The practice was continued by Roman worshipers and persisted for some time among

Christians. Other religions became opposed to castration. It is formally forbidden in both Judaism and Islam, for humans and animals alike.

Castration after puberty produces eunuchs, typically greatly reducing or completely abolishing the sex drive. Muscle mass, physical strength, and body hair may all decrease. Infertile eunuchs employed as servants or guards of harems provide the best known example of male castration to prevent cuckoldry. However, in historical times eunuchs were far more commonly employed as domestic servants, courtiers, senior political officials, military commanders, and even governors because they were less likely to foment any kind of opposition.

Intentional castration to produce eunuchs is first recorded for the Sumerian city of Lagash around 4,000 years ago. In ancient China, castration not only served as a traditional punishment but also was a requirement for entry into imperial service. At the end of the Ming Dynasty there were some 70,000 eunuchs in the service of the emperor, including some inside the Imperial Palace. Employment of eunuchs in China finally ceased in 1912. Similarly, eunuchs were employed in imperial palaces in India as messengers, attendants, and guards for female royalty. They often achieved high social standing, because some also served as royal advisors.

Castration has been variously used to make male slaves more docile, to punish criminals or enemies, and to prevent sex offenders from repeating their crimes. In *Ever Since Adam and Eve,* Malcolm Potts and Roger Short report that in the early decades of the twentieth century more than twenty states in the United States routinely castrated men on grounds of rape conviction or mental illness. Surgical or chemical castration, described as "voluntary," has been practiced in many regions, including North America and Europe. Far more rarely, human castration is used as a direct method of birth control. For instance, the Moriori population in the Chatham Islands off New Zealand castrated a proportion of male infants in order to keep the population in check after English sealers destroyed the island's seal colony, the main source of food and clothing.

Undoubtedly, the least defensible kind of castration was that applied to prepubescent European boys to prevent their voices from breaking. Boys castrated before puberty do not undergo the transformation of the larynx that would otherwise occur and therefore retain a high-pitched voice, along with weakly developed muscles, and small genitals. At the time, women were not allowed to sing in choirs in the Roman Catholic Church, so boys were castrated to preserve a special high-pitched quality of their singing. A castrato has an unusual singing voice approximately comparable to that of a soprano, mezzo-soprano, or contralto. The earliest known documentation of such castrati is contained in Italian church records from the 1550s, and castrati were part of the choir of the Sistine Chapel by 1558. In 1589, Pope Sixtus V issued the bull *Cum pro nostri temporali munere* to reorganize the choir of St Peter's Basilica in Rome, explicitly including castrati. According to one estimate, by the 1720s and 1730s, when the fad for such artificial voice preservation reached its peak, more than 4,000 boys were being castrated each year. But increasing opposition gradually led to suppression of this practice. In fact, Pope Benedict XIV made moves to ban castrati from churches in 1748, but the practice continued for another 130 years, eventually lasting for over three centuries. In 1878 Pope Leo XIII at last prohibited employment of new castrati by the Roman Catholic Church, and an official end to the practice was pronounced in 1903 by the new pope, Pius X. The last surviving Sistine castrato, Alessandro Moreschi, died in 1922.

It may seem strange to end a discussion of contraception with an account of castration. However, it is fundamentally relevant for the following reason: It beggars belief that minor interventions in human reproductive biology can provoke religious outrage, whereas the castration of young boys simply to embellish the choir (to avoid admitting women) was tolerated and even encouraged in places of worship until the beginning of the twentieth century.

LET US TURN now to the practice of assisted reproduction, where knowledge of biological processes is used to promote reproduction rather than curtail it. Proper understanding of the human menstrual cycle is vital for any attempt to treat infertility. Pioneering exploration of the female cycle by Hermann Knaus, Kyusaku Ogino, and others laid the groundwork for assisted reproduction as well as for birth control. In fact, Ogino developed his cycle-tracking method with the intention of helping couples augment their chances of conception by identifying the period of greatest fertility. Ogino opposed using his method for birth control, arguing that its failure rate was unacceptably high. He felt that promoting cycle tracking in competition with more effective contraceptive methods would result in many unwanted pregnancies terminated by abortions. How inappropriate, then, that the rhythm technique came to be known as the Ogino-Knaus method.

Infertility is a major problem. By 1985, it was estimated that in the United States alone the annual cost of treating some 2 million infertile couples was $64 billion, yet the success rate was only one in seven. Although methods have progressively improved over the past twenty-five years, low success rates continue to be a problem. Assisted reproduction undoubtedly would benefit from improved understanding of the dynamics of human conception, particularly regarding implications of timeworn sex cells and filtering mechanisms leading to miscarriages.

THE MOST STRAIGHTFORWARD method of assisted reproduction is artificial insemination (AI), in which a semen sample is introduced into the female reproductive tract to procure conception. It was Lazzaro Spallanzani, designer of those fertilization-blocking taffeta pants for frogs, who performed the first recorded successful artificial insemination in mammals. In 1784, he reported that an inseminated dog had given birth to three puppies after the usual two-month pregnancy. AI has a long history in laboratory studies and in animal husbandry, where it is now widely and routinely used for selective breeding. It was reportedly used in Arabia for horse breeding as early as the fourteenth century, although the initial history of the practice is not reliably documented.

More than a century then elapsed before accounts emerged in 1897 that AI had been successfully used in individual studies with rabbits, dogs, and horses. Among others, reproductive biologist Walter Heape—author of the term "estrus"—conducted experiments to further test its applications.

In 1785, renowned Scottish surgeon John Hunter made a major breakthrough with his first attempts at human artificial insemination, resulting in the birth of a child. However, as with domestic mammals, it took more than a century for the technique to become established for regular human use. Modern routine use of AI in humans has two basic approaches. In the first, semen is collected from a subfertile male partner of a couple that has previously failed to achieve pregnancy. The man's semen is enhanced in various ways and then transferred to his partner's womb (artificial insemination by husband, AIH). The second kind of AI, used in cases where the male partner is effectively infertile, is to use semen donated by another man of proven fertility (artificial insemination by donor, AID). With both approaches, either insemination can be timed to take place close to the natural ovulation time or ovulation can be triggered with hormone treatment. The fact that gynecologists generally prefer to induce ovulation rather than rely on it to occur naturally somewhere around midcycle is testimony to the great variability of menstrual cycles.

Despite its widespread use, we still have much to learn about underlying processes of AI. Two examples suffice to make the point. In 1984, gynecologist Esteban Kesserü reported that the likelihood of conception with sperms from a fertile donor is increased if the woman recipient copulates with her infertile partner a few hours after AI. In a similar vein, Taiwanese gynecologist Fu-Jen Huang examined pregnancy rates resulting from a male partner's sperms in a comparison between control couples treated only with AI and matched couples who followed AI with copulation around sixteen hours later. Huang and colleagues reported in 1998 that pregnancy rate was increased by follow-up copulation in couples where the male partner had a low count of motile sperms, but not if he had a high sperm count.

AI CAN BE USED when the primary cause of infertility lies with the male partner, as it does in roughly half of cases. But treatment of female infertility requires a different approach. One common cause, affecting about one in five infertile women, is blocked oviducts. Women with this condition usually have normal cycles accompanied by ovulation, but physical obstruction of sperm passage prevents fertilization. In such cases, it is now possible to retrieve egg cells from a woman's ovary, mature and fertilize them in vitro (Latin for "in the glass"), and then transfer resulting embryos to her womb. Infants born through in vitro fertilization are commonly known as "test-tube babies," although the procedure is usually carried out in a flat, shallow petri dish.

IVF must, of course, be followed by embryo transfer. Here again, it was Walter Heape who pioneered the technique, experimenting with embryo transfer in rabbits in 1890. Finally, it was John Rock who first recovered an intact fertilized egg, and he investigated IVF using sperms stored in a deep freeze. Together with Arthur Hertig and Miriam Menkin, Rock reportedly carried out the first experiments on human IVF in the United States in 1944.

The first successful human application of IVF followed by embryo transfer took place in Britain in 1977 when thirty-year-old Lesley Brown underwent treatment because of blocked oviducts. The procedure was carried out by gynecologist Patrick Steptoe and physiologist Robert Edwards, and the result was the famous pregnancy leading to the birth of Louise Brown on July 25, 1978. In 1981, Edwards reported on the rapid progress that had taken place in IVF combined with embryo transfer in just three years. At that stage, eggs were obtained by harvesting follicles from the ovary just before ovulation. In some cases, eggs were harvested by tracking a woman's normal cycles; in others, it followed targeted hormone treatment. For both, development of follicles was monitored either by measuring estrogen levels or with ultrasound. With a targeted hormonal regime, ovulation was stimulated at the appropriate time by injecting the human pregnancy hormone hCG. As long as the male partner's semen was normal, the success rate of IVF was quite high, around 90 percent. In almost all cases, only a single sperm took part in fertilization. However, the most difficult and unpredictable stage of the procedure proved to be implantation after embryo transfer. Only one in five blastocysts transferred to the womb implanted successfully.

In 2010, reproductive biologist Mark Connolly and colleagues reported that more than 3.5 million babies had been born worldwide between 1978 and 2008 using IVF and related methods of assisted reproduction. In recognition of this dramatic achievement, Edwards was belatedly awarded the Nobel Prize in Physiology or Medicine in 2010, some thirty years after his initial success.

But IVF has drawbacks. Multiple births have occurred in about one in four pregnancies, compared with only one in almost a hundred births resulting from natural conceptions. Moreover, more IVF babies are born prematurely, and perinatal mortality is almost 2 percent, double that for controls. Malformations are also somewhat higher in IVF babies. There is evidence that IVF followed by embryo transfer is associated with an increased risk of birth defects. In a large-scale study of more than 60,000 births in Ontario, Darine El-Chaar at the University of Ottawa found that the risk of birth defects for babies born through IVF was about 60 percent higher than for those born after natural conception. Similarly, a review by Jennita Reefhuis of data from the National Birth Defects Prevention Study in the United States revealed that certain birth defects were significantly more common in infants conceived with IVF. Causes of the increased risk of birth defects associated with IVF have not yet been determined. Medication used to induce ovulation is one potential cause, and some studies have indicated that the higher risk in fact reflects problems that caused infertility in the first place. Laboratory handling of eggs and sperms may play a part, and finally IVF may partially bypass natural mechanisms that would normally filter out early embryos with defects.

THE NEXT STEP in development of IVF was storing early human embryos in liquid nitrogen rather than transferring them at once to the womb. Reproductive biologist Gerard Zeilmaker and colleagues reported the first successful pregnancies obtained with frozen human embryos in 1984. By 2008 it was estimated that up to half a million IVF babies had been born from frozen embryos stored in liquid nitrogen. If IVF produces several embryos and some are not used for immediate transfer, patients may opt for long-term storage of additional embryos in liquid nitrogen. Estimates are that more than half a million frozen embryos are currently held in storage in the United States alone. Embryos resulting from fertility treatments may be donated to another woman, and eggs and sperms may be used to produce embryos that can be frozen and stored specifically for donation. But spare embryos in the freezer also raise thorny ethical and legal issues.

IVF opened up the possibility of enabling women to become pregnant after menopause. The womb is fully capable of supporting a pregnancy even after menopause and IVF has permitted women to become pregnant in their fifties or even sixties.

One crucial concern about storage of frozen early embryos is that degradation may occur. In a 2010 paper, gynecologist Ryan Riggs and colleagues examined whether storage time adversely affects survival and pregnancy outcome. He reviewed a sample of almost 12,000 frozen human embryos covering the period from 1986 to 2007 and found that storage time had no significant effect on survival after thawing. Moreover, storage duration had no significant impact on rates of clinical pregnancy, miscarriage, implantation, or live birth resulting from IVF. The primary complication of IVF using stored embryos is the risk of multiple births. This is a direct result of the dubious practice of transferring multiple embryos to raise the odds of pregnancy.

As with artificial insemination, IVF combined with embryo transfer can also be restricted to the partners of a couple or be carried out with eggs donated by another woman. It is also possible to carry out in vitro fertilization of an egg from a woman who is infertile, for instance because she has no functional womb, and then transfer the fertilized egg to the womb of a fertile woman for development. With heterosexual couples, semen from the male partner or from a donor can be used. Lesbian couples may also use the procedure with donor semen in order to have a child. The first transfer of an IVF embryo from one woman to another resulting in pregnancy was reported in July 1983, leading to the first birth in 1984. This procedure, known as surrogate motherhood, occurs in various contexts and has aroused controversy, notably because it raises complex legal issues. A surrogate mother may be paid for her services, resulting in unfortunate labels such as "wombs for rent" and "outsourced pregnancies." The procedure is illegal in some countries and officially condoned by others.

Development of IVF opened the way for fertilizing an egg with a single sperm injected by micromanipulation. In 1992, embryologists Paul Devroey and André Van Steirteghem pioneered the procedure now known as intracytoplasmic sperm injection (ICSI). ICSI is most commonly used to overcome male infertility, but it is also applied in cases where eggs are not easily penetrated by sperms. Sometimes it is simply used to ensure fertilization in vitro, especially with donor sperms. ICSI has aroused concerns because it bypasses the natural process of sperm selection during fertilization. It is now possible to test the egg-binding capacity of mature sperms to preselect them for ICSI. Nevertheless, accumulating evidence indicates that ICSI, as with IVF following simple exposure of an egg to multiple sperms, is associated with a higher incidence of birth defects. The first report on this was published in 2002 by public health specialist Michèle Hansen. Hansen and her team concluded that in infants conceived following ICSI the risk of a major birth defect is twice as high as in naturally conceived infants. However, the same result is obtained with IVF alone, so there is no indication of any added risk associated with direct injection of a single sperm. In 2004, inventors Devroey and Van Steirteghem reviewed ten years of experience with ICSI. They emphasized that as a result of this procedure, many couples can now have their own genetic child instead of having to resort to artificial insemination with donor sperms. They also confirmed that although the risk of major congenital malformation is somewhat higher with ICSI than with normal conception, the risk is no higher than with IVF alone.

IT MIGHT BE EXPECTED that religious objections would focus largely or exclusively on birth control and that interventions aimed at promoting births would be welcomed. It turns out that assisted reproduction is commonly subject to religious prohibition as well, once again on the grounds that it is "unnatural." For instance, in the 1968 circular letter *Humanae Vitae* the Roman Catholic Church opposed all kinds of artificial insemination, including IVF, because it separates childbearing from its customary marital context. In a similar vein, in the magisterial instruction *Dignitas personae*, issued in 2008, Pope Benedict XVI explicitly condemned intracytoplasmic sperm injection. There are also ethical and religious objections to destruction of early embryos generated by IVF. It is notable that the philosopher John Harris posed the following conundrum in a 2003 paper: If IVF in fact leads to less embryonic death than natural reproduction, then presumably it would be more ethical to use IVF routinely instead of natural reproduction.

The topic of intervention in human reproduction is fraught with thorny issues. However, a strong case can surely be made both for birth control, above all to tackle the runaway population explosion that affects us all, and for assisted reproduction to help childless couples in their understandable struggle to achieve parenthood. We have come a long way since our evolutionary lineage diverged from that of the most closely related species, and we have used our greatly enlarged brains to modify our lives in many different ways. Reproduction, like every other aspect of human existence, has been transformed in the process. We can no longer return to the conditions experienced by our remote gathering-and-hunting ancestors. But we can at least identify the natural basis for human reproduction and ensure that any interventions are biologically appropriate.

* * *

It is appropriate to close this final chapter with some comments on the future of human reproduction, drawing upon evolutionary evidence.

The average human ejaculate contains around a quarter of a billion sperms, just one of which can fertilize an egg. We still do not know why so many sperms are needed to accomplish the task. Successive filters along the female tract probably whittle sperms down to an elite cluster of survivors that are allowed to approach the egg. We know of at least one of these filters: Mucus in the neck of the womb blocks the passage of deformed sperms. It also seems likely that binding of sperm to the oviduct wall favors high-quality sperms. Is the rest of the journey just a random process? This is another area where more research is sorely needed. In view of the convincing evidence for declining sperm counts, we ignore this issue at our peril.

Perhaps the most worrying issue of all is genetic manipulation and the potential for human cloning. At present, certain forms of genetic manipulation are permitted in the United States and Europe under strictly regulated conditions, though both direct cloning of individuals and deliberate interference with the germ line—the special cells that produce sperms or eggs—are outlawed. Genetic manipulation of an individual while strictly avoiding the germ line seems unobjectionable. Can it be guaranteed, however, that introduction of DNA sequences into a human patient using some kind of carrier will not have unintended effects? If DNA is inserted into the human germ line by accident or design, it will automatically pass from one generation to another forever afterward. Moreover, although several countries have outlawed deliberate interference with the germ line and cloning, the technology for these procedures is readily available.

Like all other organisms, humans arose by evolution under natural conditions, and we need to explore that background to fully understand ourselves and safeguard our future. Yet the evolution of human reproduction has rarely been examined in depth, in spite of the fact that breeding success lies at the very heart of natural selection. Admittedly, reconstructing the evolutionary history of our own reproduction faces the problem that fossil evidence can only provide a few indirect pointers. But much can be achieved by studying ourselves and making comparisons with other living species, seeking general principles that will permit confident interpretation. Wide-ranging comparisons at different levels in the Tree of Life are needed to answer many basic questions about the evolution of human reproduction.

This has been a guiding beacon throughout my forty-year personal voyage of discovery to trace the natural history of our reproduction, culminating in this book. The aim throughout was to seek deeper knowledge and to identify as many practical conclusions as possible, leading me from the basic biology of sperms and eggs up to the complexities of birth control and assisted reproduction. I hope that, like me, the reader will see human reproduction in a new light.

ACKNOWLEDGMENTS

Throughout my research, the quest for clues to the evolution of human reproduction has been a major source of motivation and pleasure. Among other things, it has led me to study various primates breeding in captivity, to observe some of them in their natural habitats, to examine many museum specimens (performing occasional dissections as necessary), to pass countless hours painstakingly measuring tiny quantities of hormones, and to delve into sequences of the fundamental genetic material, DNA. But above all I have worked my way through a vast literature in a personal quest for clues to the evolutionary puzzle of human reproduction. Published information is a goldmine that yields valuable nuggets at almost every turn. Eventually, I consulted over 5,000 scientific papers and books to distill the essence into this book.

I owe an enormous debt of gratitude to many people who contributed to my education and research over the years. As a student of zoology at the University of Oxford, I was particularly inspired by lectures on animal behavior by the late Niko Tinbergen, who emphasized the natural environment as the setting in which behavior evolves. Niko lit the flame that illuminated my induction into research. My growing interest in animal behavior was also fostered by Richard Dawkins, who had just embarked on his own doctoral research in Oxford. He provided guidance and encouragement during a practical course in which I carried out my very first research project. When I graduated in 1964, my fascination with animal behavior led me to work at the Max Planck Institute for Behavioural Physiology in Seewiesen, Germany. There I conducted research on treeshrews in the mammal laboratory established by Irenäus Eibl-Eibesfeldt and benefited from many discussions with Konrad Lorenz. As it turned out, the external examiner for my doctoral thesis back in Oxford was Desmond Morris, who had just published *The Naked Ape* to great acclaim.

I owe my initiation into primate fieldwork to Jean-Jacques Petter, who supervised my postdoctoral research at the General Ecology division of the French National

Museum of Natural History in Brunoy. After initial work with his breeding colony of tiny mouse lemurs, I was able to study them in the field in 1968 while leading the Oxford Expedition to Madagascar, with support from the Royal Society of London. Some years later, the Royal Society also supported a two-year field study of bush babies in South Africa, in collaboration with Simon Bearder, a supremely talented primate fieldworker.

My academic career began in earnest in 1969, when I was appointed to a lectureship in biological anthropology at University College London. There, I benefited from the outstanding mentorship of Nigel Barnicot, a cherished colleague and friend whose early death was a bitter blow. I eventually spent thirteen years in the Department of Anthropology at University College London, with a four-year interruption while I worked as a senior research fellow in charge of the Wellcome Institute of Comparative Physiology at the Zoological Society of London. During my spell at the Zoological Society, I met up with Alan Dixson, who joined me to work on a postdoctoral project with owl monkeys. Alan and I have collaborated in one way or another ever since, and he rapidly became one of my dearest friends. While I was at the Zoological Society, together with Brian Seaton, Maya Stavy, and others I became involved in assaying reproductive hormones in urine samples from gorillas and orangutans kept in zoos. Such work on monitoring the breeding of endangered species in captivity led directly to a long-lasting relationship with the Jersey Wildlife Preservation Trust (since renamed the Durrell Wildlife Conservation Trust) and to deeply treasured friendships with honorary director Gerald Durrell and zoological director Jeremy Mallinson.

Serving as director and professor at the Anthropological Institute in Zürich, Switzerland, from 1986 to 2001 enabled me to gain additional valuable experience with primate reproduction. In particular, Christopher Pryce, a former PhD student from London, joined me as postdoctoral scientist in a research project on Goeldi's monkeys from Bolivia. Chris has been a greatly appreciated colleague and friend ever since. We established a collaborative arrangement with Max Döbeli in the nearby Veterinary Institute to conduct hormone assays, first with urine samples and then with fecal samples as well. This capacity proved to be crucial for several other studies, notably Nina Bahr's doctoral project on the relationship between hormones and maternal behavior in gorillas. I was also very fortunate in that Gustl Anzenberger, who has a special gift for care of animals in captivity, was able to join me at the Anthropological Institute. Gustl, like me, had worked at the Max Planck Institute in Seewiesen earlier in his career and became a soul mate with whom I could share interests in primate behavior.

In 2001, I became vice president and then provost of academic affairs at The Field Museum in Chicago. When the five-year contract for my administrative appointment came to an end, I was appointed as the A. Watson Armour III Curator of Biological Anthropology. This enabled me to return to full-time research. I owe a huge debt of gratitude to The Field Museum under the presidency of John McCarter, not only for enabling me to pursue my research in new directions but also for providing me with a unique opportunity to engage directly in promoting the public understanding of science. I am also particularly grateful to three close colleagues at the University of Chicago—Dario Maestripieri, Callum Ross, and Russell Tuttle—who together serve as a tremendous source of expertise and support in primatology.

In addition to those already mentioned by name, and to generations of students who have encouraged me with smart and enthusiastic responses, motivating me to try to get better every step of the way, I would like to thank the many people who contributed more directly to the development of this book. First and foremost, I owe an enormous debt to my wife, Anne Elise Martin, my constant companion and supporter. She provided invaluable insights and perceptive feedback on the manuscript at all stages. I am also very grateful to my agent, Esmond Harmsworth of Zachary Shuster Harmsworth, who expertly drew upon his considerable experience to provide me with a crash course on writing for a general audience. Albert Einstein once wisely said: "If you can't explain it simply, you don't understand it well enough." But Esmond also taught me that the story has to be made interesting; simplification and plain speaking are not enough. I also owe a debt of gratitude to my skilled editors at Basic Books, Tisse Takagi and TJ Kelleher, for shepherding me through the process of publication. Particular thanks go to John Donohue and Sue Warga for their outstandingly conscientious and skillful copyediting.

Heartfelt thanks are also due to several valiant helpers who read drafts to provide comments. Ken Kaye, an accomplished author, and Christie Henry and Glynn Meter, both skilled editors, read individual chapters and provided invaluable guidance in the art of writing. Others read the entire manuscript and helped me to improve it in various ways: my sons Oliver and Christopher Martin, my daughter, Alexandra Martin, my sister, Valerie Angus, my dear friends Marjorie Benton, Alan Dixson, and Peter Freeman, my esteemed colleague, Alaka Wali, my can-do former research assistant, Edna Davion, and interns Timothy Murphy, Lu Yao, Hannah Koch, and Andrea Rummel. I also owe particular gratitude to several interns and volunteers who hunted down obscure references to fill in the many gaps as work on the manuscript progressed: Catherine Althaus, Heather Baker, Joe Cottral, Victoria DeMartelly, Hannah Koch, Tim Murphy, Andrea Rummel, Sarah Sticha, and Meghan White. In time-honored fashion, despite this veritable squadron of helpers, I suppose that I must reluctantly accept responsibility for any errors that remain.

I also acknowledge the following former students and additional colleagues who over the years have personally contributed, however unwittingly, to the material presented: Andrew Barbour, Rosemary Bonney, Heather Brand, Bryan Carroll, Matt Cartmill, Pierre Charles-Dominique, Janice Clift, Juliette Cross, Deborah Curtis, Christopher Dean, Andrea Dettling, Frances D'Souza, Anna Feistner, Stephen Ferrari, Dirk Fleming, Michel Genoud, Paul Harvey, Michael Heistermann, Charlotte Hemelrijk, Marcel Hladik, Keith Hodges, Katherine Homewood, Louise Humphrey, Karin Isler, Mike Jerky, Alison Jolly, Susan Kingsley, the late Devra Kleiman, Lesley Knapp, Rolf Kümmerli, Elisabeth Langenegger, Ann MacLarnon, Alan McNeilly, Lara Modolo, Theya Molleson, Alexandra Müller, Thomas Mutschler, Caroline Nievergelt, Ann-Kathrin Oerke, Jennifer Pastorini, Martine Perret, William Pestle, Arlette Petter-Rousseaux, Alison Richard, Caroline Ross, Ben Rudder, Jeffrey Schwartz, Christophe Soligo, Robert Sussman, Ian Tattersall, Urs Thalmann, Krisztina Vàsàrhelyi, Franziska von Segesser, Rüdiger Wehner, Sybille Wehner, Jean Wickings, and Lesley Willner.

In closing, I would like to comment briefly on the actual writing of this book. Apart from various invited articles aimed at a general public, this is my first attempt to write for a general audience. As an academic, I have been inclined to use technical terms (although I have gradually learned to reduce them to a minimum even in the classroom), packing in illustrations and references to original publications. It has been a liberating experience to write a manuscript in which I have avoided off-putting jargon to the best of my ability and eliminated diagrams and formal references in the text. But I have tried throughout to maintain accuracy while writing plain English. I have always felt that the most admirable books for a general readership are those that manage to explain complex material without slipping into half-truths. My aim was to meet that exacting standard. Not long ago, my boredom while sitting in slow-moving traffic in Chicago was suddenly relieved by a bumper sticker on the car in front: "Eschew obfuscation." That says it all; believe me, I have tried.

GLOSSARY

Abdomen: Technical term for the belly.

Abdominal cavity: Technical name for the belly cavity.

AI: Abbreviation for **artificial insemination**.

AID: Abbreviation for **artificial insemination** by donor, used in cases where the male partner is effectively infertile.

AIH: Abbreviation for **artificial insemination** by husband.

Altricial offspring: Mammal infants that are poorly developed at birth, with the eyes and ears sealed with membranes and little or no fur. Examples: carnivores, rodents, and treeshrews.

Amenorrhea: The absence of menstruation in a woman during her reproductive years.

Amnion: One of the embryonic membranes, surrounding the embryo at first and then the fetus. The fluid contained serves to protect the fetus from mechanical shock.

Amniotic fluid: Fluid in the amnion surrounding and protecting the fetus in the womb. Rupture of the amnion at birth releases the fluid, known as the "water."

Ampulla: Expanded upper region of the **oviduct** (derived from the Latin for "flask").

Androgen: A natural or synthetic substance, usually a **steroid hormone**, that promotes or regulates the development and maintenance of male characteristics in mammals and other vertebrates.

Andrology: The branch of medicine dealing with the physiology and pathology of the male genital tract and male fertility.

Anisogamy: Difference between male and female gametes, especially in size.

Arachidonic acid (AA): One of the most important **long-chain polyunsaturated fatty acids**, particularly because of its role in brain development.

Artificial insemination: Introduction of semen into the reproductive tract of a female mammal to procure fertilization as an alternative to natural insemination. Artificial insemination in humans was at first achieved by deposition of semen in the vagina or oviduct, but it has become increasingly common to inject appropriately processed semen directly into the womb (**intrauterine insemination**).

Asexual reproduction: An alternative name for **clonal reproduction**.

Aspermia: A complete lack of semen.

Assisted reproduction: Any procedure used to combat infertility, including **artificial insemination**, **in vitro fertilization**, **intracytoplasmic sperm injection**, and **embryo transfer**.

Atresia: Degeneration of a **follicle** in the **ovary**.

Azoospermia: Absence of sperm cells in the semen.

Baculum: Penis bone (also known as the **os penis**).

Barrier contraception: Any form of contraception using a physical obstacle to prevent fertilization, such as diaphragms or condoms.

Basal body temperature (BBT): The lowest body temperature while resting, usually during sleep. In women, a small rise in temperature by between 0.5°F and 1°F commonly occurs just after ovulation. The raised level is usually maintained for the rest of the cycle, during the **luteal phase**. The change in BBT over the cycle closely tracks a parallel change in a woman's energy turnover.

Blastocyst: A structure formed from the fertilized egg in the early development of any vertebrate. In humans, formation of the blastocyst begins on the fifth day after fertilization. It has an inner cell mass, which goes on to form the embryo, and an outer layer of cells, the **trophoblast**, which contributes to formation of the placenta after **implantation**. There are about eighty cells in the human blastocyst.

Breech birth: Condition in which the baby enters the birth canal with its buttocks or feet leading, in contrast to the normal headfirst presentation. Breech birth is slower than headfirst delivery.

Caesarean delivery: Extraction of the baby, placenta, and membranes through incisions in the walls of the mother's abdomen and womb.

Casein: Name given to special proteins found in the milk of all mammals. Caseins make up four-fifths of the protein content of cow's milk and around a third of the proteins in human milk.

Central nervous system (CNS): The brain and spinal cord.

Cervical mucus: Mucus produced by the neck of the womb. Monitoring of cervical mucus is often used in women to detect the time of **ovulation**, when it typically has a particularly thin and watery consistency.

Cervix: Technical name for the neck of the womb.

Chorionic gonadotropin: A hormone produced during pregnancy after implantation of the **blastocyst**. It is subsequently produced by the placenta until birth occurs. The form produced during human pregnancy is known as **human chorionic gonadotropin** (**hCG**). In chemical structure, chorionic gonadotropin is closely similar to **luteinizing hormone** (**LH**).

Chromosomes: Structures in the cell nucleus made up of the genetic material DNA and various proteins. Most of an animal's genetic information is encoded in the DNA in chromosomes, although some information is also encoded in **mitochondria**.

Chronobiology: The study of biological rhythms (cycles) in living organisms and their entrainment to ambient light cues.

Circadian clock: An internal clock that governs biological processes over each twenty-four-hour period (from the Latin *circa* for "approximately" and *dies* for "day"). The clock has a basic duration of about twenty-four hours and is adjusted (entrained) each day according to changing light levels.

Circannual clock: An internal clock that governs biological processes over the course of the year (from the Latin *circa* for "approximately" and *annus* for "year"). The clock has a basic duration of about twelve months and is adjusted (entrained) by environmental cues, notably day length (which changes in a predictable fashion over the annual cycle).

Clonal reproduction: Propagation by self-replication rather than through sexual reproduction.

CNS: Abbreviation for the **central nervous system**.

Colostrum: Special kind of milk produced by a mother just after birth, usually starting within one day and lasting three to four days. Colostrum contains antibodies to protect the newborn infant against disease. It typically contains less fat and more protein than milk produced subsequently.

Condom: A barrier device often used during coitus as a means of contraception. Usually refers to a sheath fitted over the erect penis, but female condoms are also made. Apart from blocking fertilization, condoms also reduce the transmission of sexual diseases.

Contraception: Intentional avoidance of conception by using various sexual practices, devices, chemical compounds, or surgery. Also known as birth control or family planning. Widely used methods include barrier techniques that stop sperms from reaching the egg (e.g., **condoms**, **diaphragms**), killing of sperm (**spermicides**), inhibition of **ovulation** (oral contraceptive pills), prevention of implantation (**intrauterine devices**), and eliminating sperms from the seminal fluid (**vasectomy**). Sexual practices include the **rhythm method** and coitus interruptus.

Contraction: Intermittent, painful tightening of the womb muscles during labor, reducing the volume of the uterus and pushing the fetus toward the birth canal.

Contracture: Mild tightening of the womb muscles. Contractures occur throughout pregnancy and switch to more powerful **contractions** during labor.

Corpus luteum: "Yellow body" formed from the remains of the ovarian follicle after **ovulation**, typically producing **progesterone** in response to **luteinizing hormone**.

Cortisol: A steroid hormone produced by the adrenal gland. It is released in response to stressors and is commonly measured as an index of stress. Its main functions are to increase blood sugar levels, to suppress the immune system, and to promote metabolism of fats, proteins, and carbohydrates.

Cryptorchidism: Failure of one or both testes to descend. Once testes are formed in the human fetus they stay inside the abdomen until about the seventh month of pregnancy. In about 97 percent of cases the testicles are in the scrotum before birth. About 80 percent of retained testes descend during the first year of life, mostly within three months. So the true incidence of cryptorchidism is around 1 percent overall and about 0.15 percent for failure of both testes to descend.

Diaphragm: A dome-like barrier device, made of soft latex or silicone, that is fitted over the entry to the neck of the womb as a means of birth control. The rim contains a spring so that it presses against the walls of the vagina to prevent passage of semen.

Docosahexaenoic acid (DHA): One of the most important **long-chain polyunsaturated fatty acids**, notably because of its role in brain development.

Eclampsia: A rare, sometimes fatal outcome of **preeclampsia**, characterized by convulsions.

Ectopic pregnancy: Attachment and development of an embryo in the wrong place (oviduct or abdominal cavity).

Embryo: In mammals, the embryonic stage is the initial period during which different tissues are developed and the basic framework of the offspring's body slowly develops. It begins with conception, passes through implantation and then continues on through the initial period of exchange via the placenta. In human development, the embryonic stage lasts eight weeks after conception and is then followed by the fetal stage.

Embryology: The biological study of early growth and development of animals and plants.

Embryo transfer: A step in the process of assisted reproduction with humans or other mammals in which one or more embryos are placed into a female's womb. Regularly used in combination with **in vitro fertilization** (IVF).

Endometrium: The inner lining of the womb.

Epidemiology: The branch of medicine dealing with patterns of health, disease, and associated factors at the population level.

Epididymis: A coiled, elongated duct extending down the posterior border of the testis in a male mammal. It serves for storage, maturation, and transit of sperms and leads into the **vas deferens**. The length of the human epididymis is approximately twenty feet, and sperms take twelve to twenty-one days to move through it.

Episiotomy: An incision made to enlarge the vulva during birth.

Estrogen: A natural or synthetic substance, usually a **steroid hormone**, that promotes or regulates the development and maintenance of female characteristics in mammals and other vertebrates. The three main estrogens in mammals are estradiol, estriol, and estrone.

Estrus: A recurring state of sexual excitation in a female mammal, when she is most receptive to mating. Commonly known as "heat" or "rut." Estrus typically occurs just before ovulation when the female is able to conceive. Among mammals, estrus is lacking only in the higher primates (monkeys, apes, and humans).

Extrapair copulation: Any copulation that occurs outside of a mated adult pair of animals.

Extrapair paternity: Offspring sired outside of a mated adult pair of animals.

Fallopian tube: Alternative name for the **oviduct**.

Fecundity: The biological capacity to conceive.

Fertility: Achievement of pregnancy.

Fetus: The term used for the developing offspring in the womb once individual major organ systems of the body—brain, heart, gut, and urogenital apparatus—can be recognized. Unlike an **embryo**, a fetus broadly resembles a newborn individual and differs mainly in being small. In human development, the fetal stage lasts about thirty weeks, starting eight weeks after conception and ending with birth.

First stage of labor: Stage beginning with regular, painful contractions and ending when the neck of the womb is fully open.

Follicular phase: First half of the **ovarian cycle**, from menstruation to ovulation in higher primates. The **ovarian follicle** ripens during this phase.

FSH: Follicle-stimulating hormone, produced by the **pituitary gland**. As its name indicates, this hormone promotes growth of **ovarian follicles**.

Full-term birth: Any human birth following a pregnancy lasting between thirty-seven and forty-two weeks after the onset of the last menstrual period.

Gamete intrafallopian transfer (GIFT): A form of assisted reproduction in which eggs are collected from a woman's ovaries and then placed in one of the **oviducts**, along with a sperm sample. This allows fertilization to take place at the natural site.

Gametes: Sex cells; eggs in females and sperms in males.

Genome: A complete set of hereditary information, in most cases consisting of DNA. Cells with a proper nucleus have two genomes: a nuclear genome contained in the chromosomes and a separate genome in each **mitochondrion**.

Germ line: A separate, everlasting lineage of founder cells that produce sex cells in many-celled organisms.

Gestation period: Strictly speaking, the interval between conception and birth in mammals generally. In humans, the true gestation period is about two weeks shorter than the pregnancy length calculated from the onset of the last menstrual period.

GIFT: Abbreviation for **gamete intrafallopian transfer**.

Graafian follicle: See **ovarian follicle**.

Gynecology: The medical-surgical specialty dealing with the female reproductive organs in their nonpregnant state.

hCG: Human **chorionic gonadotropin**.

Hermaphrodite: An individual with both male and female reproductive organs.

Higher primates: Monkeys, apes, and humans, which are generally more advanced than **lower primates** in features such as the brain. Also known as **simians**.

Hydatiform mole: Abnormality of pregnancy in which exclusive presence of paternal chromosomes results in a huge placenta and virtually no embryo.

Hysterectomy: Surgical removal of the womb.

ICSI: Abbreviation for **intracytoplasmic sperm injection**.

Implantation: A process early in pregnancy in which the **blastocyst** becomes attached to the wall of the womb. In humans and great apes, the blastocyst actually becomes embedded in the womb lining.

Induced ovulation: Ovulation that is triggered by the act of mating.

Infertility: Inability to reproduce. Currently defined for humans by the World Health Organization as "a failure to conceive after at least 12 months of unprotected intercourse."

Inguinal canal: The canal through which the testis passes to leave the abdominal cavity and enter the scrotal sac.

Intracytoplasmic sperm injection (ICSI): A procedure used in **in vitro fertilization** in which a single sperm is injected directly into an egg.

Intrauterine device (IUD): A form of birth control in which a small device, often containing copper, is inserted into the womb. This is a long-term, reversible form of **contraception**. Although the IUD has popularly been called a "coil," it is now more usually T-shaped.

Intrauterine insemination: A form of **artificial insemination** in which appropriately processed semen is injected directly into the womb.

In vitro fertilization (IVF): Fertilization of an egg by a sperm outside the body in a glass container (in vitro). Infants born as a result of IVF are commonly known as "test-tube babies," although fertilization usually takes place in a flat petri dish. IVF is used to treat infertility in cases where other methods of **assisted reproduction** have been unsuccessful.

Isthmus: Narrow lower region of the **oviduct** (derived from the Greek *isthmos*, referring to a narrow passage).

IUD: Abbreviation for **intrauterine device**.

IVF: Abbreviation for **in vitro fertilization**.

Lactation: The production of milk to feed offspring; a defining characteristic of mammals.

Laparoscope: A narrow tube that is inserted through an incision in the wall of the abdomen and used to view internal organs.

LCPUFA: Abbreviation for **long-chain polyunsaturated fatty acid**.

LH: Abbreviation for **luteinizing hormone**.

Long-chain polyunsaturated fatty acid (LCPUFA): A fatty acid is long-chained if it contains a string of more than eighteen carbon atoms, and it is polyunsaturated if there are two or more double bonds between adjacent carbon atoms. Simply stated, this means that LCPUFAs can add chemical links at sites where there are double bonds. Two very important LCPUFAs, notably found in human milk, are **arachidonic acid** and **docosahexaenoic acid**.

Lower primates: Lemurs, lorises, and tarsiers, which generally have remained more primitive than **higher primates** in features such as the brain. Also known as **prosimians**.

Luteal phase: Second half of the **ovarian cycle**, from ovulation to the next menstruation in higher primates. After ovulation, the ruptured **ovarian follicle** forms a **corpus luteum** that persists during the second half of the cycle.

Luteinizing hormone (LH): One of the hormones produced by the **pituitary gland**. A spike in LH generally seems to trigger ovulation in mammals. LH also promotes growth of the **corpus luteum** in the ovary and influences the testis in males.

Marsupial: A member of the subclass of mammals known as Marsupialia, which have short pregnancies and give birth to very small offspring whose subsequent development commonly occurs in a pouch (*marsupium* in Latin).

Mastitis: Inflammation of the breast, commonly due to infection.

Melatonin: A hormone secreted by the **pineal gland** in the brain in mammals. Known as the "hormone of darkness," it is released only during the night and is directly involved in coordinating biological clocks.

Menarche: The onset of menstruation at puberty in girls (from the Greek *meno-*, "month," and *arkhe*, "beginning").

Menopause: Permanent cessation of reproduction in women, occurring at about fifty years of age. From the Greek *meno-*, "month," and *pausis*, "cessation."

Menstrual cycle: The name given to the **ovarian cycle** in **higher primates** and a few other mammals, which ends with **menstruation**.

Menstruation: Periodic shedding of the inner lining of the womb accompanied by loss of blood at the end of the **ovarian cycle**.

Miscarriage: Pregnancy loss during the first eighteen weeks after human conception. Technically known as **spontaneous abortion**.

Mitochondrion (pl. **mitochondria**): A small inclusion enclosed by a membrane found in most cells that have a nucleus. Mitochondria were originally free-living bacteria, related to the bacteria that cause typhus, that became permanent residents in cells or other organisms and now function as tiny "powerhouses."

Monogamy: A mating or marriage system in which one adult male associates with a single adult female.

Monotreme: An egg-laying mammal (platypus or echidna) belonging to the mammal subclass Monotremata.

Multigravida: A woman who has had two or more pregnancies.

Multimale group: A social group containing two or more adult males along with two or more adult females.

Multiparous: Adjective applied to females who have had more than one pregnancy.

Nägele's rule: A standard way of calculating the due date for a pregnancy. The rule estimates the expected date of delivery by adding one year to the first day of a woman's last menstrual period, subtracting three months, and adding seven days. The result is approximately 280 days (forty weeks) from the last menstrual period.

Neuron: The technical name for a nerve cell.

New World monkeys: Monkeys living in South and Central America.

Obstetrics: The medical-surgical specialty dealing with the female reproductive organs in their pregnant state.

Old World monkeys: Monkeys living in Africa, Asia, and Southeast Asia.

One-male group: A social group containing a single adult male along with one or more adult females.

Oocyte: Ripe egg cell at the time of ovulation.

Oogenesis: The process of production of eggs in the ovary.

Oogonium (pl. **oogonia**)**:** A primordial starter-cell in the ovary of a female fetus that can give rise to an **oocyte**.

Oophorectomy: Technical term for female castration.

Oosik: Alaskan name for the penis bone of an adult male walrus.

Orchidectomy: Technical term for male castration.

Os penis: Penis bone (also known as a **baculum**).

Osteoporosis: Reduction in bone density.

Ovarian bursa: A membranous pouch that typically encloses the ovary in mammals, apart from tarsiers and **higher primates**.

Ovarian cycle: The cyclical process in the ovary of a female mammal in which **ovulation** can take place. The part of the cycle before ovulation is the **follicular phase**, while the part after ovulation is the **luteal phase**.

Ovarian follicle: Structure containing the developing egg.

Ovariectomy: Surgical removal of one or both ovaries. Can be used as a form of female castration.

Ovary: The egg-producing organ in the reproductive system of a female vertebrate, typically present as a pair, with one on the left side and one on the right side. In addition to releasing eggs, the ovary also produces hormones.

Oviduct: The tube through which an egg passes from the ovary to the womb in mammals. Also known as the **Fallopian tube**.

Ovipary: Reproduction by laying eggs, with embryos developing largely or completely outside the mother's body. Many vertebrates (most fish, amphibians, and reptiles; all birds; monotremes) and most invertebrates (e.g., insects and mollusks) are oviparous.

Ovulation: Release of an egg from the ovary.

Ovum: The technical name for an egg.

Oxytocin: A hormone produced by the pituitary gland that is best known for its involvement in reproduction of mammals, especially during and after birth. Large quantities of oxytocin are released after distension of the neck of the womb during labor, facilitating birth, maternal bonding, and suckling (after the teats have been stimulated). Oxytocin is sometimes called the "love hormone" because it also plays a part in orgasm, pair bonding, and maternal behavior.

Parental care: Any form of parental behavior that increases offspring fitness.

Parental investment: Any investment by the parent in an individual offspring that increases the survival and reproductive success of that offspring at a cost to the parent's ability to invest in other offspring, or any character or action that increases offspring fitness at a cost to any component of parental fitness.

Parturition: Technical term for birth.

Paternity certainty: The average probability that a male is the sire of a given offspring or set of offspring.

Pediatrician: A physician specializing in care of infants and children.

Perineum: The anatomical region between the vulva and the anus in a female mammal and between the scrotum and the anus in a male.

Pessary: A device for insertion into the vagina, either as a support for the uterus or (**diaphragm pessary**) to deliver a drug, such as a contraceptive.

Pineal gland: This tiny gland in the brain in mammals is a remnant of a third eye that was once present in the skull roof of early reptiles. It secretes the hormone **melatonin**, which plays a direct part in coordinating biological clocks.

Pituitary gland: A small, hormone-producing gland on the underside of the brain, described as the conductor of the body's hormonal orchestra. Notable for producing the reproductive hormones **FSH**, **LH**, and **prolactin**.

Placenta: The structure attaching the embryo/fetus to the wall of the womb and serving for supply of nutrients and removal of waste products. The name is derived from the Greek *plakous*, referring to a flat cake. The human placenta is also known as the afterbirth.

Placenta previa: An abnormality of pregnancy in which the placenta is located over the neck of the womb rather than at its upper end. Because the placenta is dislodged as birth approaches, severe bleeding can occur and delivery is usually carried out with Caesarian section.

Placentophagy: Eating the placenta following birth. Mothers of most mammal species show this behavior.

Polyandry: A mating or marriage system in which one adult female associates with two or more adult males.

Polygyny: A mating or marriage system in which one adult male associates with two or more adult females.

Polyzoospermy: A condition in which the ejaculate contains excessive sperm numbers, with a density greater than 250 million per milliliter of semen or a total sperm count of more than 800 million in men.

Postnatal: Following birth.

Postpartum depression: A nonpsychotic depressive episode that begins in or extends into the period after birth.

Precocial offspring: Mammal infants that are well developed at birth, with the eyes and ears open and a good coat of fur. Examples: hoofed mammals and primates.

Preeclampsia: An adverse condition of pregnancy characterized by high blood pressure, elimination of protein in the urine, and swelling due to water retention (edema). If symptoms worsen, full-blown eclampsia may ensue.

Preformationism: The obsolete notion that an organism arises through enlargement of a miniature version of itself rather than gradually developing its parts through transformation from a fertilized egg. Spermists proposed that a tiny homunculus is present in the head of the sperm, while ovists postulated that a female equivalent is contained in the egg.

Premature baby: A human baby born more than three weeks before the due date, after a pregnancy of thirty-seven weeks or less, dated from the first day of the last menstrual period.

Prenatal: Before birth.

Preterm birth: Any human birth delivered at less than thirty-seven completed weeks of pregnancy as dated from the first day of the last menstrual period.

Primigravida: A woman experiencing her first pregnancy.

Primiparous: Adjective applied to females who have had only one pregnancy.

Progesterone: A **steroid hormone** that serves important functions both in the ovarian cycle and during pregnancy. It is typically produced by the **corpus luteum** after ovulation, during the **luteal phase** of the cycle. If conception occurs, it is then directly involved in the maintenance of pregnancy until birth.

Progestogen: A chemical substance, usually a **steroid**, that acts like the hormone **progesterone**.

Prolactin: A hormone produced and released by the **pituitary gland**, connected with **lactation** in mammals. It also has functions in paternal behavior and in relation to stress.

Prosimian: Any member of the "**lower primates**," which include all lemurs, lorises, and tarsiers.

Prostaglandins: Locally acting chemical messengers derived from fatty acids. They have various effects, such as controlling contraction and relaxation of smooth muscle. They are not hormones as they are produced in many locations throughout the body and not at one specific site. Prostaglandins were first discovered in seminal fluid and were so named because they were thought to originate from the **prostate gland**. It is now known that they were actually derived from the **seminal vesicles**.

Prostate gland: A walnut-sized gland of the reproductive system located between the bladder and the penis in most male mammals, including humans. The prostate secretes a milky fluid that makes up about a third of the volume of the semen in humans, together with sperms and fluid from the **seminal vesicles**.

Pseudocyesis: Technical name for **pseudopregnancy**.

Pseudopregnancy: False or phantom pregnancy. A condition resembling pregnancy that occurs in some female mammals. It is often caused by presence of a corpus luteum following infertile copulation and occasionally by tumors that disrupt the normal hormonal balance. Pseudopregnancy is common in dogs, for example, but rare in women.

Relaxin: A hormone produced in females by the **corpus luteum** of the ovary, by the **placenta**, and by the breast. In males, it is produced by the **prostate gland** and is present in **semen**. At the time of birth, relaxin serves to soften the ligament joining the two halves of the pelvis in the pubic region, but it also has other functions.

Rhythm method: A method of family planning involving periodic abstinence; often described as "natural." The basic approach is to identify a fertile window in a woman's menstrual cycle during which conception may occur and to confine coitus to other days in the cycle. As characteristics of cycles differ between women, individual record keeping is recommended in order to recognize a personal average (calendar method). The method has been further refined by recording of **basal body temperature** and monitoring the condition of **cervical mucus**.

Scrotal sac: A sac containing a testis outside the main body cavity.

Scrotum: A pouch of skin containing the testes outside the main body cavity.

Seasonal affective disorder (SAD): A medical condition in which episodes of depression occur at a particular time of the year. SAD mainly occurs during winter but may arise at other times of the year.

Second stage of labor: Stage beginning when the neck of the womb is fully open and ending with the birth of the baby.

Semen: The milky white fluid that is ejaculated from the penis, consisting of sperms carried in the **seminal fluid**.

Semenogelin: A prominent protein in semen that is involved in coagulation after ejaculation.

Seminal fluid: The sperm-carrying fluid in mammals, containing the sugar fructose. The fluid is mainly secreted by the **seminal vesicles**, but secretion from the prostate gland also contributes.

Seminal vesicles: A pair of simple tubular glands located in the pelvic cavity behind and below the urinary bladder in male mammals, including humans. The seminal vesicles secrete most of the fluid in **semen**. Birds also have structures called seminal vesicles, but they serve to store sperm.

Sex cells: Eggs and sperms, produced by a separate **germ line** in many-celled organisms.

Sexual dimorphism: A physical difference between adult males and females of the same species with no direct connection to the primary differences in sex organs. Dimorphism most commonly occurs in body size and general appearance.

SIDS: Abbreviation for **sudden infant death syndrome**.

Simian: Any member of the "**higher primates**," which include all monkeys, apes, and humans.

Spermatogenesis: The process through which sperms are produced in the testis. In humans, production of sperms takes just over two months from start to finish.

Spermatozoon (pl. **spermatozoa**): The technical name for a sperm (from the ancient Greek *sperma*, "seed," and *zōon*, "living thing").

Spermicide: A chemical compound that kills sperms.

Spontaneous abortion: Technical name for pregnancy loss during the first eighteen weeks after conception (miscarriage). Miscarriage during the first month after conception is regarded as early pregnancy loss; miscarriage more than a month after conception is clinical spontaneous abortion.

Spontaneous generation: The once widely held notion that living organisms can arise directly from nonliving matter; for example, mice from stored grain or fly maggots from dead meat.

Spontaneous ovulation: Ovulation that takes place without the need for mating. It is triggered by an internal spike in **luteinizing hormone (LH)** produced by the pituitary gland. Ovulation in humans, as in other primates, is spontaneous.

Steroid hormone: Any member of the steroid class of organic compounds that acts as a hormone. Steroid hormones with a prominent sexual function are androgens, estrogens, and progestogens.

Sudden infant death syndrome (SIDS): The sudden, unexpected death of an infant that remains unexplained after a thorough medical investigation. Because the greatest risk for SIDS is while babies are asleep, it is also known as cot death or crib death.

Testis (pl. **testes**): The paired male organ responsible for sperm production and the release of certain hormones, notably **testosterone**.

Testosterone: The best known **androgen**, a **steroid hormone** that promotes or regulates the development and maintenance of male characteristics in mammals and other vertebrates.

Third stage of labor: Stage beginning with the birth of the baby and ending with the expulsion of the placenta.

Trophoblast: The outer layer of the **blastocyst**, formed soon after fertilization, which participates in **implantation** and development of the **placenta**.

Tubal ligation: A method of sterilization involving severing, cauterizing, or clamping both oviducts.

Ultrasound scan: A painless examination using sound waves to create images of organs and other structures inside the body. It is routinely used to examine development of the fetus and may also be employed for special purposes such as monitoring of **ovulation**.

Urology: The branch of medicine concerned with the urinary tracts of both men and women and with the entire reproductive system of men.

Uterus: Technical name for the womb.

Vagina: Section of the female reproductive tract serving for intromission of the penis.

Varicocele: Abnormal enlargement of the vein in the scrotum that drains blood from the testes. Varicoceles occur in 15 percent of men and are presumed to be an evolutionary consequence of our upright body posture. Most varicoceles (more than 80 percent) occur on the left side and the remainder on both sides. They generally arise during puberty. About a third of men evaluated for infertility have a varicocele (twice the normal incidence).

Vas deferens (pl. **vasa deferentia**): Part of the reproductive system in many male vertebrates. A duct that transports stored sperm from the tail of the testicular **epididymis** to the penis when ejaculation occurs. In men, each vas deferens is about twelve inches long.

Vasectomy: A method of male sterilization involving severing of the **vas deferens**.

Vivipary: Reproduction through development of the embryo inside the mother's body, leading to live birth, rather than by laying eggs. Among vertebrates, all placental mammals and marsupials are viviparous, as are a minority of fish, amphibians, and reptiles.

Vulva: The external opening of a woman's reproductive tract.

REFERENCES

BOOKS

Abitbol, M. M., F. A. Chervenah, and W. J. Ledger. 1996. *Birth and Human Evolution: Anatomical and Obstetrical Mechanisms in Primates.* Westport, CT: Bergin and Garvey.

Allman, J. 1999. *Evolving Brains.* New York: W. H. Freeman/Scientific American.

Asdell, S. A. 1946. *Patterns of Mammalian Reproduction.* London: Constable.

Baker, R. R. 2006. *Sperm Wars: Infidelity, Sexual Conflict and Other Bedroom Battles.* New York: Basic Books.

Baker, R. R., and M. A. Bellis. 1994. *Human Sperm Competition: Copulation, Masturbation and Infidelity.* London: Chapman & Hall.

Bancroft, J. 2009. *Human Sexuality and Its Problems.* 3rd ed. Edinburgh: Churchill Livingstone, Elsevier.

Beischer, N. A., E. V. Mackay, and P. B. Colditz. 1997. *Obstetrics and the Newborn: An Illustrated Textbook.* 3rd ed. London: Baillière Tindall.

Betzig, L. L., M. Borgerhoff Mulder, and P. Turke. 1988. *Human Reproductive Behaviour: A Darwinian Perspective.* Cambridge: Cambridge University Press.

Billings, J. J. 1983. *The Ovulation Method: The Achievement or Avoidance of Pregnancy by a Technique Which Is Reliable and Universally Acceptable.* Melbourne: Advocate Press.

Birkhead, T. R. 2000. *Promiscuity: An Evolutionary History of Sperm Competition and Sexual Conflict.* London: Faber and Faber.

Blum, D. 2002. *Love at Goon Park: Harry Harlow and the Science of Affection.* New York: Basic Books.

Boucke, L. 2008. *Infant Potty Training: A Gentle and Primeval Method Adapted to Modern Living.* 3rd ed. Lafayette, CO: White-Boucke Publishing.

Cobb, M. 2006. *The Egg and Sperm Race: The Seventeenth-Century Scientists Who Unravelled the Secrets of Sex, Life and Growth.* London: Free Press.

Cunnane, S. C. 2005. *Survival of the Fattest: The Key to Human Brain Evolution.* Hackensack, NJ: World Scientific.

De Jonge, C. J., and C. L. R. Barratt. 2002. *Assisted Reproductive Technology: Accomplishments and New Horizons.* Cambridge: Cambridge University Press.

Dettwyler, K., and P. Stuart-Macadam, eds. 1995. *Breastfeeding: Biocultural Perspectives.* Piscataway, NJ: Aldine Transaction.

Diamond, J. M. 1997. *Why Is Sex Fun? The Evolution of Human Sexuality.* New York: Basic Books.

Dixson, A. F. 2009. *Sexual Selection and the Origins of Human Mating Systems.* Oxford: Oxford University Press.

———. 2012. *Primate Sexuality: Comparative Studies of the Prosimians, Monkeys, Apes and Human Beings.* 2nd ed. Oxford: Oxford University Press.

Djerassi, C. 2001. *This Man's Pill: Reflections on the 50th Birthday of the Pill.* Oxford: Oxford University Press.

Edwards, R. G. 1980. *Conception in the Human Female.* London: Academic Press.

Ellison, P. T. 2001. *On Fertile Ground: A Natural History of Human Reproduction.* Cambridge, MA: Harvard University Press.

Fildes, V. A. 1986. *Breasts, Bottles and Babies: A History of Infant Feeding.* Edinburgh: Edinburgh University Press.

Ford, C. S. 1945. *A Comparative Study of Human Reproduction.* New Haven, CT: Yale University Press.

Ford, C. S., and F. A. Beach. 1951. *Patterns of Sexual Behaviour.* New York: Harper & Bros.

Gould, S. J. 1996. *The Mismeasure of Man.* 2nd ed. New York: W. W. Norton & Co.

Grosser, O. 1909. *Vergleichende Anatomie und Entwicklungsgeschichte der Eihäute und der Placenta.* Vienna: Wilhelm Braumüller.

Hartman, C. G. 1962. *Science and the Safe Period: A Compendium of Human Reproduction.* Baltimore: Williams and Wilkins Co.

Hellin, D. 1895. *Die Ursache der Multiparität der uniparen Thiere überhaupt und der Zwillingsschwangerschaft beim Menschen insbesondere.* Munich: Seitz & Schauer.

Hrdy, S. B. 2009. *Mothers and Others: The Evolutionary Origins of Mutual Understanding.* Cambridge, MA: Belknap Press of Harvard University Press.

Huntington, E. 1938. *The Season of Birth.* New York: John Wiley.

Jelliffe, D. B., E. F. P. Jelliffe, and L. Kersey. 1989. *Human Milk in the Modern World.* Oxford: Oxford University Press.

Jerison, H. J. 1973. *Evolution of the Brain and Intelligence.* New York: Academic Press.

Jirásek, J. E. 2001. *An Atlas of the Human Embryo and Fetus: A Photographic Review of Human Prenatal Development*. Boca Raton, FL: Parthenon Publishing.

Jolly, A. 1999. *Lucy's Legacy: Sex and Intelligence in Human Evolution*. Cambridge, MA: Harvard University Press.

Jordan, B. 1992. *Birth in Four Cultures: A Crosscultural Investigation of Childbirth in Yucatan, Holland, Sweden, and the United States*. 4th ed. Prospect Heights, IL: Waveland Press.

Kaye, K. 1982. *The Mental and Social Life of Babies: How Parents Create Persons*. Chicago: University of Chicago Press.

Klaus, M. H., and P. H. Klaus. 2000. *Your Amazing Newborn*. Cambridge, MA: Da Capo.

Konner, M. 2010. *The Evolution of Childhood: Relationships, Emotion, Mind*. Cambridge, MA: Belknap Press of Harvard University Press.

Latz, L. J. 1932. *The Rhythm of Sterility and Fertility in Women: A Discussion of the Physiological, Practical, and Ethical Aspects of the Discoveries of Drs. K. Ogino (Japan) and Prof. H. Knaus (Austria) Regarding the Periods When Conception Is Impossible and When Possible*. Chicago: Latz Foundation.

Leakey, M. D. 1984. *Disclosing the Past*. London: Weidenfeld & Nicolson.

Low, B. S. 2000. *Why Sex Matters: A Darwinian Look at Human Behavior*. Princeton, NJ: Princeton University Press.

Malthus, T. R. 1798. *An Essay on the Principles of Population as It Affects the Future Improvement of Society*. London: J. Johnson.

Marantz Henig, R. 2004. *Pandora's Baby: How the First Test Tube Babies Sparked the Reproductive Revolution*. Boston: Houghton Mifflin.

Marshall, J. 1963. *The Infertile Period—Principles and Practice*. London: Darton, Longman and Todd.

Martin, R. D. 1990. *Primate Origins and Evolution: A Phylogenetic Reconstruction*. Princeton, NJ: Princeton University Press.

Masters, W. H., and V. E. Johnson. 1966. *Human Sexual Response*. London: Churchill.

McLaren, A. 1992. *A History of Contraception: From Antiquity to the Present Day*. Oxford: Blackwell.

Michael, R. T., J. H. Gagnon, B. O. Laumann, and G. Kolata. 1994. *Sex in America: A Definitive Survey*. New York: Little, Brown.

Miller, G. F. 2000. *The Mating Mind: How Sexual Choice Shaped the Evolution of Human Nature*. London: Heinemann.

Morris, D. 1967. *The Naked Ape: A Zoologist's Study of the Human Animal*. London: Jonathan Cape.

Nesse, R. M., and G. C. Williams. 1995. *Why We Get Sick: The New Science of Darwinian Medicine*. New York: Times Books.

Ogino, K. 1934. *Conception Period of Women*. Harrisburg, PA: Medical Arts.

Paterniti, M. 2000. *Driving Mr. Albert: A Trip Across America with Einstein's Brain*. New York: The Dial Press.

Pinto-Correia, C. 1997. *The Ovary of Eve: Eggs and Sperm and Preformation*. Chicago: University of Chicago Press.

Pollard, I. 1994. *A Guide to Reproduction: Social Issues and Human Concerns*. Cambridge: Cambridge University Press.

Pond, C. M. 1998. *The Fats of Life*. Cambridge: Cambridge University Press.

Portmann, A. 1990. *A Zoologist Looks at Human Kind*. New York: Columbia University Press.

Potts, M., and R. V. Short. 1999. *Ever Since Adam and Eve: The Evolution of Human Sexuality*. Cambridge: Cambridge University Press.

Profet, M. 1997. *Pregnancy Sickness: Using Your Body's Natural Defenses to Protect Your Baby-to-Be*. Reading, MA: Perseus.

Quetelet, A. 1869. *Physique sociale ou essai sur le développement des facultés de l'homme*. Paris: J.-B. Baillière et Fils.

Redshaw, M. E., R. P. A. Rivers, and D. B. Rosenblatt. 1985. *Born Too Early: Special Care for Your Preterm Baby*. Oxford: Oxford University Press.

Riddle, J. M. 1992. *Contraception and Abortion from the Ancient World to the Renaissance*. Cambridge, MA: Harvard University Press.

Robin, P. 1998. *When Breastfeeding Is Not an Option: A Reassuring Guide for Loving Parents*. Roseville, CA: Prima Lifestyles.

Rock, J. C. 1963. *The Time Has Come: A Catholic Doctor's Proposals to End the Battle over Birth Control*. New York: Knopf.

Shields, B. 2005. *Down Came the Rain: My Journey Through Postpartum Depression*. New York: Hyperion.

Small, M. 1998. *Our Babies, Ourselves: How Biology and Culture Shape the Way We Parent*. New York: Anchor Books.

Smolensky, M. H., and L. Lamberg. 2000. *The Body Clock Guide to Better Health: How to Use Your Body's Natural Clock to Fight Illness and Achieve Maximum Health*. New York: Henry Holt.

Symons, D. 1979. *The Evolution of Human Sexuality*. Oxford: Oxford University Press.

Tanner, J. M. 1989. *Foetus into Man: Growth from Conception to Maturity*. 2nd ed. Hertfordshire, UK: Castlemead.

Taylor, G. 2000. *Castration: An Abbreviated History of Western Manhood*. London: Routledge.

Thornhill, R., and S. W. Gangestad. 2008. *The Evolutionary Biology of Human Female Sexuality*. Oxford: Oxford University Press.

Tone, A. 2001. *Devices and Desires: A History of Contraception in America.* New York: Hill and Wang.

Trevathan, W. R. 1987. *Human Birth: An Evolutionary Perspective.* Hawthorne, NY: Aldine de Gruyter.

Vollman, R. F. 1977. *The Menstrual Cycle.* Philadelphia: W. B. Saunders.

Wolf, J. H. 2001. *Don't Kill Your Baby: Public Health and the Decline of Breastfeeding in the Nineteenth and Twentieth Centuries.* Columbus: Ohio State University Press.

Wood, J. W. 1995. *Dynamics of Human Reproduction: Biology, Biometry, Demography.* New York: Aldine de Gruyter.

World Health Organization. 2003. *Global Strategy for Infant and Young Child Feeding.* Geneva: WHO Press.

———. 2005. *Guiding Principles for Feeding Non-Breastfed Children 6–24 Months.* Geneva: WHO Press.

———. 2006. *Pregnancy, Childbirth, Postpartum and Newborn Care: A Guide for Essential Practice.* 2nd ed. Geneva: WHO Press.

———. 2010. *WHO Laboratory Manual for the Examination and Processing of Human Semen.* 5th ed. Geneva: WHO Press.

Worth, J. 2002. *Call the Midwife: A True Story of the East End in the 1950s.* Twickenham, UK: Merton Books.

Wrangham, R. 2009. *Catching Fire: How Cooking Made Us Human.* New York: Basic Books.

SCIENTIFIC ARTICLES

Abou-Saleh, M. T., R. Ghubash, L. Karim, M. Krymski, and I. Bhai. 1998. Hormonal Aspects of Postpartum Depression. *Psychoneuroendocrinology* 23: 465–475.

Ahlfeld, F. 1869. Beobachtungen über die Dauer der Schwangerschaft. *Monatschr Geburtsh Frauenkrankh* 34: 180–225.

Aitken, R. J., P. Koopman, and S. E. M. Lewis. 2004. Seeds of Concern. *Nature* 432: 48–52.

Albers, L. L. 1999. The Duration of Labor in Healthy Women. *J Perinatol* 19: 114–119.

Alliende, M. E. 2002. Mean Versus Individual Hormonal Profiles in the Menstrual Cycle. *Fertil Steril* 78: 90–95.

Allsworth, J. E., J. Clarke, J. F. Peipert, M. R. Hebert, A. Cooper, and L. A. Boardman. 2007. The Influence of Stress on the Menstrual Cycle Among Newly Incarcerated Women. *Wom Health Iss* 17: 202–209.

Altmann, J., and A. Samuels. 1992. Costs of Maternal Care: Infant-carrying in Baboons. *Behav Ecol Sociobiol* 29: 391–398.

Anderson, J. W., B. M. Johnstone, and D. T. Remley. 1999. Breast-feeding and Cognitive Development: A Meta-analysis. *Am J Clin Nutr* 70: 525–535.

Anderson, K. G. 2006. How Well Does Paternity Confidence Match Actual Paternity? *Curr Anthropol* 47: 513–520.

Anderson, M. J., and A. F. Dixson. 2002. Motility and the Midpiece in Primates. *Nature* 416: 496.

Anderson, M. J., J. Nyholt, and A. F. Dixson. 2005. Sperm Competition and the Evolution of Sperm Midpiece Volume in Mammals. *J Zool Lond* 267: 135–142.

Andrade, A. T. L., J. P. Souza, S. T. Shaw, E. M. Belsey, and P. J. Rowe. 1991. Menstrual Blood Loss and Iron Stores in Brazilian Women. *Contraception* 43: 241–249.

Auger, J., J. M. Kunstmann, F. Gzyglik, and P. Jouannet. 1995. Decline in Semen Quality Among Fertile Men in Paris During the Past Twenty Years. *New Engl J Med* 332: 281–285.

Backe, B. 1991. A Circadian Variation in the Observed Duration of Labor. *Acta Obstet Gynecol Scand* 70: 465–468.

Baker, T. G. 1963. A Quantitative and Cytological Study of Germ Cells in Human Ovaries. *Proc R Soc Lond B Biol Sci* 158: 417–433.

Ben Shaul, D. M. 1962. The Composition of the Milk of Wild Animals. *Int Zoo Yearb* 4: 333–342.

Benshoof, L., and R. Thornhill. 1979. The Evolution of Monogamy and Concealed Ovulation in Humans. *J Soc Biol Struct* 2: 95–106.

Bergsjø, P., D. W. Denman, H. J. Hoffman, and O. Meirik. 1990. Duration of Human Singleton Pregnancy: A Population-Based Study. *Acta Obstet Gynecol Scand* 69: 197–207.

Bernier, M. O., G. Plu-Bureau, N. Bossard, L. Ayzac, and J. C. Thalabard. 2000. Breastfeeding and Risk of Breast Cancer: A Meta-analysis of Published Studies. *Hum Reprod Update* 6: 374–386.

Bielert, C., and J. G. Vandenbergh. 1981. Seasonal Influences on Births and Male Sex Skin Coloration in Rhesus Monkeys (*Macaca mulatta*) in the Southern Hemisphere. *J Reprod Fertil* 62: 229–233.

Billings, J. J. 1981. Cervical Mucus: The Biological Marker of Fertility and Infertility. *Int J Fertil* 26: 182–195.

Birch, E. E., S. Garfield, D. R. Hoffman, R. Uauy, and D. G. Birch. 2000. A Randomized Controlled Trial of Early Dietary Supply of Long-Chain Polyunsaturated Fatty Acids and Mental Development in Term Infants. *Dev Med Child Neurol* 42: 174–181.

Bogin, B. 1997. Evolutionary Hypotheses for Human Childhood. *Yearb Phys Anthropol* 40: 63–89.

Boklage, C. E. 1990. Survival Probability of Human Conceptions from Fertilisation to Term. *Int J Fertil* 35: 75–94.

Bonde, J. P., E. Ernst, T. K. Jensen, N. H. Hjollund, H. Kolstad, T. B. Henriksen, T. Scheike, A. Giwercman, J. Olsen, and N. E. Skakkebaek. 1998. Relation Between Semen Quality and Fertility: A Population-Based Study of 430 First-Pregnancy Planners. *Lancet* 352: 1172–1177.

Bostofte, E., J. Serup, and H. Rebbe. 1982. Relation Between Sperm Count and Semen Volume, and Pregnancies Obtained During a Twenty-Year Follow-Up Period. *Int J Androl* 5: 267–275.

Bovens, L. 2006. The Rhythm Method and Embryonic Death. *J Med Ethics* 32: 355–356.

Boyle, P., S. N. Kaye, and A. G. Robertson. 1987. Changes in Testicular Cancer in Scotland. *Eur J Cancer Clin Oncol* 23: 827–830.

Bronson, F. H. 1995. Seasonal Variation in Human Reproduction: Environmental Factors. *Quart Rev Biol* 70: 141–164.

Brophy, J. T., M. M. Keith, A. Watterson, R. Park, M. Gilbertson, E. Maticka-Tyndale, M. Beck, H. Abu-Zahra, K. Schneider, A. Reinhartz, R. DeMatteo, and I. Luginaah. 2012. Breast Cancer Risk in Relation to Occupations with Exposure to Carcinogens and Endocrine Disruptors: A Canadian Case-Control Study. *Environm Health* 11, 87.

Brosens, J. J., M. G. Parker, A. McIndoe, R. Pijnenborg, and I. A. Brosens. 2009. A Role for Menstruation in Preconditioning the Uterus for Successful Pregnancy. *Am J Obstet Gynecol* 200(6): 615.e1–6.

Brown, J. E., E. S. Kahn, and T. J. Hartman. 1997. Profet, Profits, and Proof: Do Nausea and Vomiting of Early Pregnancy Protect Women from "Harmful" Vegetables? *Am J Obstet Gynecol* 176: 179–181.

Brummelte, S., and L. A. M. Galea. 2010. Depression During Pregnancy and Postpartum: Contribution of Stress and Ovarian Hormones. *Prog Neuro-Psychopharmacol Biol Psychiatry* 34: 766–776.

Buckley, S. J. 2006. Placenta Rituals and Folklore from Around the World. *Midwifery Today Int Midwife* 80: 58–59.

Bujan, L., M. Daudin, J.-P. Charlet, P. Thonneau, and R. Mieusset. 2000. Increase in Scrotal Temperature in Car Drivers. *Hum Reprod* 15: 1355–1357.

Burley, N. 1979. The Evolution of Concealed Ovulation. *Am Nat* 114: 835–858.

Burr, M. L., E. S. Limb, M. J. Maguire, L. Amarah, B. A. Eldridge, J. C. Layzell, and T. G. Merrett. 1993. Infant Feeding, Wheezing, and Allergy: A Prospective Study. *Arch Dis Childh* 68: 724–728.

Byard, R. W., M. Makrides, M. Need, M. A. Neumann, and R. A. Gibson. 1995. Sudden Infant Death Syndrome: Effect of Breast and Formula Feeding on Frontal Cortex and Brainstem Lipid Composition. *J Paediatr Child Health* 31: 14–16.

Cancho-Candela, R., J. M. Andres-de Llano, and J. Ardura-Fernandez. 2007. Decline and Loss of Birth Seasonality in Spain: Analysis of 33 421 731 Births over 60 Years. *J Epidemiol Commun Health* 61: 713–718.

Carlsen, E., A. Giwercman, N. Keiding, and N. E. Skakkebaek. 1992. Evidence for Decreasing Quality of Semen During Past 50 Years. *Brit Med J* 305: 609–613.

Carnahan, S. J., and M. I. Jensen-Seaman. 2008. Hominoid Seminal Protein Evolution and Ancestral Mating Behavior. *Am J Primatol* 70: 939–948.

Caro, T. M. 1987. Human Breasts, Unsupported Hypotheses Reviewed. *Hum Evol* 2: 271–282.

Carpenter, C. R. 1942a. Sexual Behavior of Free Ranging Rhesus Monkeys (*Macaca mulatta*). I. Specimens, Procedures and Behavior Characteristics of Estrus. *J Comp Psychol* 33: 113–142.

———. 1942b. Sexual Behavior of Free Ranging Rhesus Monkeys (*Macaca mulatta*). II. Periodicity of Estrus, Homosexual, Autoerotic and Non-conformist Behavior. *J Comp Psychol* 33: 143–162.

Chandwani, K. D., I. Cech, M. H. Smolensky, K. Burau, and R. C. Hermida. 2004. Annual Pattern of Human Conception in the State of Texas. *Chronobiol Int* 21: 73–93.

Chard, T. 1991. Frequency of Implantation and Early-pregnancy Loss in Natural Cycles. *Baillière's Clin Obstet Gynaecol* 5: 179–189.

Chauhan, S. P., J. A. Scardo, E. Hayes, A. Z. Abuhamad, and V. Berghella. 2010. Twins: Prevalence, Problems, and Preterm Births. *Am J Obstet Gynecol* 203: 305–315.

Chen, A. M., and W. J. Rogan. 2004. Breastfeeding and the Risk of Postneonatal Death in the United States. *Pediatrics* 113: e435–e439.

Chilvers, C., M. C. Pike, D. Forman, K. Fogelman, and M. E. J. Wadsworth. 1984. Apparent Doubling of Frequency of Undescended Testis in England and Wales in 1962–81. *Lancet* 324: 330–332.

Chua, S., S. Arulkumaran, I. Lim, N. Selamat, and S. S. Ratnam. 1994. Influence of Breastfeeding and Nipple Stimulation on Postpartum Uterine Activity. *Brit J Obstet Gynaecol* 101: 804–805.

Clark, N. L., and W. J. Swanson. 2005. Pervasive Adaptive Evolution in Primate Seminal Proteins. *PLoS Genet* 1(3): e35.

Collaborative Group on Hormonal Factors in Breast Cancer. 1996. Breast Cancer and Hormonal Contraceptives: Collaborative Reanalysis of Individual Data on

53 297 Women With Breast Cancer and 100 239 Women Without Breast Cancer from 54 Epidemiological Studies. *Lancet* 347: 1713–1727.

———. 2002. Breast Cancer and Breastfeeding: Collaborative Reanalysis of Individual Data from 47 Epidemiological Studies in 30 Countries, Including 50,302 Women With Breast Cancer and 96,973 Women Without the Disease. *Lancet* 360: 187–196.

Conaway, C. H., and C. B. Koford. 1964. Estrous Cycles and Mating Behavior in a Free-Ranging Band of Rhesus Monkeys. *J Mammal* 45: 577–588.

Conaway, C. H., and D. S. Sade. 1965. The Seasonal Spermatogenic Cycle in Free Ranging Rhesus Monkeys. *Folia Primatol* 3: 1–12.

Connolly, M. P., S. Hoorens, and G. M. Chambers. 2010. The Costs and Consequences of Assisted Reproductive Technology: An Economic Perspective. *Hum Reprod Update* 16: 603–613.

Consensus Statement: Breastfeeding as a Family Planning Method. 1988. *Lancet* 332: 1204–1205.

Cooper, T. G., E. Noonan, S. von Eckardstein, J. Auger, H. W. G. Baker, H. M. Behre, T. B. Haugen, T. Kruger, C. Wang, M. T. Mbizvo, and K. M. Vogelsong. 2010. World Health Organization Reference Values for Human Semen Characteristics. *Hum Reprod Update* 16: 231–245.

Coqueugniot, H., J.-J. Hublin, F. Veillon, F. Houët, and T. Jacob. 2004. Early Brain Growth in *Homo erectus* and Implications for Cognitive Ability. *Nature* 431: 299–302.

Cowgill, U. M. 1966a. Historical Study of the Season of Birth in the City of York, England. *Nature* 209: 1067–1070.

———. 1966b. The Season of Birth in Man. *Man n.s.* 1: 232–241.

———. 1966c. Season of Birth in Man: Contemporary Situation with Special Reference to Europe and the Southern Hemisphere. *Ecology* 47: 614–623.

Cunnane, S. C., and M. A. Crawford. 2003. Survival of the Fattest: Fat Babies Were the Key to Evolution of the Large Human Brain. *Comp Biochem Physiol A* 136: 17–26.

Cunningham, A. S., D. B. Jelliffe, and E. F. P. Jelliffe. 1991. Breastfeeding and Health in the 1980s: A Global Epidemiological Review. *J Pediatr* 118: 659–666.

Czeizel, A. E., E. Puho, N. Acs, and F. Banhidy. 2006. Inverse Association Between Severe Nausea and Vomiting in Pregnancy and Some Congenital Abnormalities. *Am J Med Genet* 140A, 453–462.

Danilenko, K., and E. A. Samoilova. 2007. Stimulatory Effect of Morning Bright Light on Reproductive Hormones and Ovulation: Results of a Controlled Crossover Trial. *PLoS Clin Trials* 2(2): e7.

de Boer, C. H. 1972. Transport of Particulate Matter Through the Female Genital Tract. *J Reprod Fertil* 28: 295–297.

DeSilva, J. M. 2011. A Shift Toward Birthing Relatively Large Infants Early in Human Evolution. *Proc Natl Acad Sci USA* 108: 1022–1027.

Dettwyler, K. A. 2004. When to Wean: Biological Versus Cultural Perspectives. *Clin Obstet Gynecol* 47: 712–723.

deVries, M. W., and M. R. deVries. 1977. Cultural Relativity of Toilet Training Readiness: A Perspective from East Africa. *Pediatrics* 60: 170–177.

Dixson, B. J., G. M. Grimshaw, W. L. Linklater, and A. F. Dixson. 2010. Eye-tracking of Men's Preferences for Waist-to-Hip Ratio and Breast Size of Women. *Arch Sex Behav* 40: 43–50.

Djerassi, C., and S. P. Leibo. 1994. A New Look at Male Contraception. *Nature* 370: 11–12.

Dodds, E. C., and W. Lawson. 1936. Synthetic Estrogenic Agents Without the Phenanthrene Nucleus. *Nature* 137: 996.

———. 1938. Molecular Structure in Relation to Oestrogenic Activity. Compounds Without a Phenanthrene Nucleus. *Proc Roy Soc Lond B* 125: 222–232.

Dorus, S., P. D. Evans, G. J. Wyckoff, S. S. Choi, and B. T. Lahn. 2004. Rate of Molecular Evolution of the Seminal Protein Gene SEMG2 Correlates with Levels of Female Promiscuity. *Nature Genet* 36: 1326–1329.

Doyle, R. 1996. World Birth-Control Use. *Sci Am* 275(9): 34.

Dunn, P. M. 2000. Dr. Emmett Holt (1855–1924) and the Foundation of North American Paediatrics. *Arch Dis Child Fetal Neonatal Ed* 83: F221–F223.

Dupras, T. L., H. P. Schwarcz, and S. I. Fairgrieve. 2001. Infant Feeding and Weaning Practices in Roman Egypt. *Am J Phys Anthropol* 115: 204–212.

Dyroff, R. 1939. Beiträge zur Frage der physiologischen Sterilität. *Zentralbl Gynäkol* 1939: 1717–1721.

Edwards, C. A., and A. M. Parrett. 2002. Intestinal Flora During the First Months of Life: New Perspectives. *Brit J Nutr* 88, S1: s11–s18.

Edwards, R. G. 1981. Test-Tube Babies, 1981. *Nature* 293: 253–256.

Egli, G. E., and M. Newton. 1961. The Transport of Carbon Particles in the Human Female Reproductive Tract. *Fertil Steril* 12: 151–155.

Eiben, B., I. Bartels, S. Bähr-Porsch, S. Borgmann, G. Gatz, G. Gellert, R. Goebel, W. Hammans, M. Hentemann, R. Osmers, R. Rauskolb, and I. Hansmann. 1990. Cytogenetic Analysis of 750 Spontaneous Abortions with the Direct-Preparation Method of Chorionic Villi and Its Implications for Studying Genetic Causes of Pregnancy Wastage. *Am J Hum Genet* 47: 656–663.

El-Chaar, D., O. Y. Yang, J. Bottomely, S. W. Wen, and M. Walker. 2006. Risk of Birth Defects in Pregnancies Associated with Assisted Reproductive Technology. *Am J Obstet Gynecol* 195: S21.

Ellington, J. E., D. P. Evenson, R. W. Wright, A. E. Jones, C. S. Schneider, G. A. Hiss, and R. S. Brisbois. 1999. Higher-Quality Human Sperm in a Sample Selectively Attach to Oviduct (Fallopian Tube) Epithelial Cells in Vitro. *Fertil Steril* 71: 924–929.

Emera, D., R. Romero, and G. Wagner. 2012. The Evolution of Menstruation: A New Model for Genetic Assimilation. *BioEssays* 34: 26–35.

Evans, K. M., and V. J. Adams. 2010. Proportion of Litters of Purebred Dogs Born by Caesarean Section. *J Small Anim Pract* 51: 113–118.

Falk, H. C., and S. A. Kaufman. 1950. What Constitutes a Normal Semen? *Fertil Steril* 1: 489–503.

Figà-Talamanca, I., C. Cini, G. C. Varricchio, F. Dondero, L. Gandini, A. Lenzi, F. Lombardo, L. Angelucci, R. Di Grezia, and F. R. Patacchioli. 1996. Effects of Prolonged Autovehicle Driving on Male Reproductive Function: A Study Among Taxi Drivers. *Am J Industr Med* 30: 750–758.

Finn, C. A. 1998. Menstruation: A Non-adaptive Consequence of Uterine Evolution. *Quart Rev Biol* 73: 163–173.

Fisch, H., E. T. Goluboff, J. H, Olson, J. Feldshuh, S. J. Broder, and D. H. Barad. 1996. Semen Analyses in 1,283 Men from the United States over a 25-Year Period: No Decline in Quality. *Fertil Steril* 65: 1009–1014.

Flaxman, S. M., and P. W. Sherman. 2000. Morning Sickness: A Mechanism for Protecting Mother and Embryo. *Quart Rev Biol* 75: 113–148.

Fleming, A. S., D. Ruble, H. Krieger, and P. Y. Wong. 1997. Hormonal and Experiential Correlates of Maternal Responsiveness During Pregnancy and the Puerperium in Human Mothers. *Horm Behav* 31: 145–158.

Foote, R. H. 2002. The History of Artificial Insemination: Selected Notes and Notables. *J Anim Sci* 80: 1–10.

Fox, C. A., S. J. Meldum, and B. W. Watson. 1973. Continuous Measurement by Radio-Telemetry of Vaginal pH During Human Coitus. *J Reprod Fertil* 33: 69–75.

Francis, C. M., E. L. P. Anthony, J. A. Brunton, and T. H. Kunz. 1994. Lactation in Male Fruit Bats. *Nature* 367: 691–692.

Franciscus, R. G. 2009. When Did the Modern Human Pattern of Childbirth Arise? New Insights from an Old Neandertal Pelvis. *Proc Natl Acad Sci USA* 106: 9125–9126.

Fuller, B. T., J. L. Fuller, D. A. Harris, and R. E. M. Hedges. 2006. Detection of Breastfeeding and Weaning in Modern Human Infants with Carbon and Nitrogen Stable Isotope Ratios. *Am J Phys Anthropol* 129: 279–293.

Galloway, T., R. Cipelli, J. Guralnik, L. Ferrucci, S. Bandinelli, A. M. Corsi, C. Money, P. McCormack, and D. Melzer. 2010. Daily Bisphenol A Excretion and Associations with Sex Hormone Concentrations: Results from the InCHIANTI Adult Population Study. *Environm Health Perspect* 118: 1603–1608.

Garwicz, M., M. Christensson, and E. Psouni. 2009. A Unifying Model for Timing of Walking Onset in Humans and Other Mammals. *Proc Natl Acad Sci USA* 106: 21889–21893.

German, J. 1968. Mongolism, Delayed Fertilization and Human Sexual Behaviour. *Nature* 217: 516–518.

Gibbons, A. 2008. The Birth of Childhood. *Science* 322: 1040–1043.

Gibson, J. R., and T. McKeown. 1950. Observations on All Births (23,970) in Birmingham, 1947. I: Duration of Gestation. *Brit J Soc Med* 4: 221–233.

———. 1952. Observations on All Births (23,970) in Birmingham, 1947. VI: Birth Weight, Duration of Gestation and Survival Related to Sex. *Brit J Soc Med* 6: 152–158.

Gilbert, S. F., and Z. Zevit. 2001. Congenital Human Baculum Deficiency: The Generative Bone of Genesis 2: 21–23. *Am J Med Genet* 101: 284–285.

Glasier, A., and A. S. McNeilly. 1990. Physiology of Lactation. *Clin Endocrinol Metab* 4: 379–395.

Goldman, A. S. 2002. Evolution of the Mammary Gland Defense System and the Ontogeny of the Immune System. *J Mammary Gland Biol Neoplas* 7: 277–289.

Goldwater, P. N. 2011. A Perspective on SIDS Pathogenesis. The Hypotheses: Plausibility and Evidence. *BMC Med* 9(64): 1–13.

Gomendio, M., and E. R. S. Roldan. 1993. Co-evolution Between Male Ejaculates and Female Reproductive Biology in Eutherian Mammals. *Proc R Soc Lond B Biol Sci* 252: 7–12.

Gould, J. E., J. W. Overstreet, and F. W. Hanson. 1984. Assessment of Human Sperm Function after Recovery from the Female Reproductive Tract. *Biol Reprod* 31: 888–894.

Gray, J. P., and L. D. Wolfe. 1983. Human Female Sexual Cycles and the Concealment of Ovulation Problem. *J Soc Biol Struct* 6: 345–352.

Gray, L., L. W. Miller, B. L. Philipp, and E. M. Blass. 2002. Breastfeeding Is Analgesic in Healthy Newborns. *Pediatrics* 109: 590–593.

Groer, M. W., M. W. Davis, and J. Hemphill. 2002. Postpartum Stress: Current Concepts and the Possible Protective Role of Breastfeeding. *J Obstet Gynecol Neonat Nurs* 31: 411–417.

Guerrero, R., and C. A. Lanctot. 1970. Aging of Fertilizing Gametes and Spontaneous Abortion: Effect of the Day of Ovulation and the Time of Insemination. *Am J Obstet Gynecol* 107: 263–267.

Guerrero, V., and O. I. Rojas. 1975. Spontaneous Abortion and Aging of Human Ova and Spermatozoa. *New Engl J Med* 293: 573–575.

Gunz, P., S. Neubauer, B. Maureille, and J.-J. Hublin. 2010. Brain Development After Birth Differs Between Neanderthals and Modern Humans. *Curr Biol* 20: R921–R922.

Guzick, D. S., J. W. Overstreet, P. Factor-Litvak, C. K. Brazil, S. T. Nakajima, C. Coutifaris, S. A. Carson, P. Cisneros, M. P. Steinkampf, J. A. Hill, D. Xu, and D. L. Vogel. 2001. Sperm Morphology, Motility, and Concentration in Fertile and Infertile Men. *New Engl J Med* 345: 1388–1393.

Häger, R. M., A. K. Daltveit, D. Hofoss, S. T. Nilsen, T. Kolaas, P. Oian, and T. Henriksen. 2004. Complications of Cesarean Deliveries: Rates and Risk Factors. *Am J Obstet Gynecol* 190: 428–434.

Haimov-Kochman, R., R. Har-Nir, E. Ein-Mor, V. Ben-Shoshan, C. Greenfield, I. Eldar, Y. Bdolah, and A. Hurwitz. 2012. Is the Quality of Donated Semen Deteriorating? Findings from a 15 Year Longitudinal Analysis of Weekly Sperm Samples. *Isr Med Assoc J* 14: 372–377.

Hallberg, L., A.-M. Hogdahl, L. Nilsson, and G. Rybo. 1966. Menstrual Blood Loss—A Population Study. *Acta Obstet Gynecol Scand* 45: 320–351.

Hammes, L. M., and A. E. Treloar. 1970. Gestational Interval from Vital Records. *Am J Public Health* 60: 1496–1505.

Hansen, M., J. J. Kurinczuk, C. Bower, and S. Webb. 2002. The Risk of Major Birth Defects After Intracytoplasmic Sperm Injection and In Vitro Fertilization. *New Engl J Med* 346: 725–730.

Hansen, M., J. J. Kurinczuk, N. de Klerk, P. Burton, and C. Bower. 2012. Assisted Reproductive Technology and Major Birth Defects in Western Australia. *Obst Gynecol* 120: 852–863.

Harcourt, A. H., P. H. Harvey, S. G. Larson, and R. V. Short. 1981. Testis Weight, Body Weight and Breeding System in Primates. *Nature* 293: 55–57.

Harcourt, A. H., A. Purvis, and L. Liles. 1995. Sperm Competition: Mating System, Not Breeding Season, Affects Testes Size of Primates. *Funct Ecol* 9: 468–476.

Harder, T., R. Bergmann, G. Kallischnigg, and A. Plagemann. 2005. Duration of Breastfeeding and Risk of Overweight: A Meta-analysis. *Am J Epidemiol* 162: 397–403.

Harlow, H. F., and M. K. Harlow. 1962. Social Deprivation in Monkeys. *Sci Am* 207(5): 136–146.

———. 1966. Learning to Love. *Am Sci* 54: 244–272.

Hartman, C. G. 1931. The Phylogeny of Menstruation. *J Am Med Ass* 97: 1863–1865.

———. 1932. Studies in the Reproduction of the Monkey *Macacus* (*Pithecus*) *rhesus*, with Special Reference to Menstruation and Pregnancy. *Contrib Embryol Carnegie Inst Wash* 23: 1–161.

Heape, W. 1900. The "Sexual Season" of Mammals and the Relation of the "Prooestrum" to Menstruation. *Quart J Micr Sci* 44: 1–70.

Hedges, L. V., and A. Nowell. 1995. Sex Differences in Mental Test Scores, Variability, and Numbers of High-Scoring Individuals. *Science* 269: 41–45.

Heikkilä, K., A. Sacker, Y. Kelly, M. J. Renfrew, and M. A. Quigley. 2011. Breast Feeding and Child Behaviour in the Millennium Cohort Study. *Arch Dis Childh* 96: 635–642.

Heird, W. C. 2001. The Role of Polyunsaturated Fatty Acids in Term and Preterm Infants and Breastfeeding Mothers. *Pediatr Clin N Am* 48: 173–188.

Helland, I. B., L. Smith, K. Saarem, O. D. Saugstad, and C. A. Drevon. 2003. Maternal Supplementation with Very-Long-Chain n-3 Fatty Acids During Pregnancy and Lactation Augments Children's IQ at 4 Years of Age. *Pediatrics* 111: e39–e44.

Heres, M. H. G., M. Pel, M. Borkent-Polet, P. E. Treffers, and M. Mirmiran. 2000. The Hour of Birth: Comparisons of Circadian Pattern Between Women Cared for by Midwives and Obstetricians. *Midwifery* 16: 173–176.

Higham, J. P., C. Ross, Y. Warren, M. Heistermann, and A. M. MacLarnon. 2007. Reduced Reproductive Function in Wild Baboons (*Papio hamadryas anubis*) Related to Natural Consumption of the African Black Plum (*Vitex doniana*). *Horm Behav* 52: 384–390.

Hill, S. A. 1888. The Life Statistics of an Indian Province. *Nature* 38: 245–250.

Hinde, K., and L. A. Milligan. 2011. Primate Milk: Proximate Mechanisms and Ultimate Perspectives. *Evol Anthropol* 20: 9–23.

Hirata, S., K. Fuwa, K. Sugama, K. Kusunoki, and H. Takeshita. 2011. Mechanism of Birth in Chimpanzees: Humans Are Not Unique Among Primates. *Biol Lett* 7: 686–688.

Holdcroft, A., A. Oatridge, J. V. Hajnal, and G. M. Bydder. 1997. Changes in Brain Size in Normal Human Pregnancy. *J Physiol* 499P: 79P–80P.

Holt, L. E. 1890. Observations upon the Capacity of the Stomach in Infancy. *Arch Pediatr* 7: 960–967.

Honnebier, M. B. O. M. 1994. The Role of the Circadian System During Pregnancy and Labor in Monkey and Man. *Acta Obstet Gynecol Scand* 73: 85–88.

Honnebier, M. B. O. M., and P. W. Nathanielsz. 1994. Primate Parturition and the Role of Maternal Circadian System. *Eur J Obstet Gynaecol Reprod Biol* 55: 193–203.

Hook, E. B., and S. Harlap. 1979. Difference in Maternal-Age Specific Rates of Down's Syndrome Between Jews of European Origin and of North African or Asian Origin. *Teratology* 20: 243.

Howie, P. W., J. S. Forsyth, S. A. Ogston, A. Clark, and C. D. Florey. 1990. Protective Effect of Breast Feeding Against Infection. *Brit Med J* 300: 11–16.

Howie, P. W., and A. S. McNeilly. 1982. Effect of Breast-feeding Patterns on Human Birth Intervals. *J Reprod Fertil* 65: 545–557.

Huang, F. J., S. Y. Chang, F. T. Kung, J. F. Wu, and M. Y. Tsai. 1998. Timed Intercourse After Intrauterine Insemination for Treatment of Infertility. *Eur J Obstet Gynecol Reprod Biol* 80: 257–261.

Hubrecht, A. A. W. 1898. Über die Entwicklung der Placenta von *Tarsius* und *Tupaia*, nebst Bemerkungen über deren Bedeutung als haemopoeitische Organe. *Proc Int Congr Zool* 4: 345–411.

Huyghe, E., T. Matsuda, and P. Thonneau. 2003. Increasing Incidence of Testicular Cancer Worldwide: A Review. *J Urol* 170: 5–11.

Iffy, L. 1963a. Embryonic Studies of Time of Conception in Ectopic Pregnancy and First Trimester Abortion. *Obstet Gynecol* 26: 490–498.

Iffy, L. 1963b. The Time of Conception in Foetal Monstrosities. *Gynaecologia* 156: 140–142.

Iffy, L., and M. B. Wingate. 1970. Risks of Rhythm Method of Birth Control. *J Reprod Med* 5: 11–17.

Insler, V., M. Glezerman, L. Zeidel, D. Bernstein, and N. Misgav. 1980. Sperm Storage in the Human Cervix: A Quantitative Study. *Fertil Steril* 33: 288–293.

Irvine, S., E. Cawood, and D. Richardson. 1996. Evidence of Deteriorating Semen Quality in the United Kingdom: Birth Cohort Study in 577 Men in Scotland over 11 Years. *Brit Med J* 312: 467–471.

Itan, Y., A. Powell, M. A. Beaumont, J. Burger, and M. G. Thomas. 2009. The Origins of Lactase Persistence in Europe. *PLoS Comput Biol* 5(8): e1000491; doi:10.1371/journal.pcbi.1000491.

James, W. H. 1971. The Distribution of Coitus Within the Human Inter-Menstruum. *J Biosoc Sci* 3: 159–171.

———. 1980. Secular Trend in Reported Sperm Counts. *Andrologia* 12: 381–388.

———. 1990. Seasonal Variation in Human Births. *J Biosoc Sci* 22: 113–119.

———. 1996. Down Syndrome and Natural Family Planning. *Am J Med Genet* 66: 365.

Jarnfelt-Samsioe, A. 1987. Nausea and Vomiting in Pregnancy: A Review. *Obstet Gynecol Surv* 41: 422–427.

Jenny, E. 1933. Tagesperiodische Einflüsse auf Geburt und Tod. *Schweiz med Wochenschr* 63: 15–17.

Jensen, T. K., N. Jørgensen, M. Punab, T. B. Haugen, J. Suominen, B. Zilaitiene, A. Horte, A.-G. Andersen, E. Carlsen, Ø. Magnus, V. Matulevicius, I. Nermoen, M. Vierula, N. Keiding, J. Toppari, and N. E. Skakkebaek. 2004. Association of In Utero Exposure to Maternal Smoking with Reduced Semen Quality and Testis Size in Adulthood: A Cross-Sectional Study of 1,770 Young Men from the General Population in Five European Countries. *Am J Epidemiol* 159: 49–58.

Jensen-Seaman, M. I., and W.-H. Li. 2003. Evolution of the Hominoid Semenogelin Genes, the Major Proteins of Ejaculated Semen. *J Mol Evol* 57: 261–270.

Jöchle, W. 1973. Coitus-Induced Ovulation. *Contraception* 7: 523–564.

Jolly, A. 1972. Hour of Birth in Primates and Man. *Folia Primatol* 18: 108–121.

———. 1973. Primate Birth Hour. *Int Zoo Yearb* 13: 391–397.

Jongbloet, P. H. 1985. The Ageing Gamete in Relation to Birth Control Failures and Down Syndrome. *Europ J Pediatr* 144: 343–347.

Jongbloet, P. H., A. J. M. Poestkoke, A. J. H. Hamers, and J. H. J. van Erkelens-Zwets. 1978. Down Syndrome and Religious Groups. *Lancet* 312: 1310.

Jørgensen, N., A.-G. Andersen, F. Eustache, D. S. Irvine, J. Suominen, J. H. Petersen, J. Holm, A. N. Andersen, A. Nyboe, J. Auger, E. H. H. Cawood, A. Horte, T. K. Jensen, P. Jouannet, N. Keiding, M. Vierula, J. Toppari, and N. E. Skakkebaek. 2001. Regional Differences in Semen Quality in Europe. *Hum Reprod* 16: 1012–1019.

Jørgensen, N., M. Vierula, R. Jacobsen, E. Pukkala, A. Perheentupa, H. E. Virtanen, N. E. Skakkebaek, and J. Toppari. 2010. Recent Adverse Trends in Semen Quality and Testis Cancer Incidence Among Finnish Men. *Int J Androl* 34: e37–e48.

Juberg, R. C. 1983. Origin of Chromosome Abnormalities: Evidence for Delayed Fertilization in Meiotic Nondisjunction. *Hum Genet* 64: 122–127.

Kaiser, I. H., and F. Halberg. 1962. Circadian Periodic Aspects of Birth. *Ann NY Acad Sci* 98: 1056–1068.

Kakar, D. N., S. Chopra, S. A. Samuel, and K. Singar. 1989. Beliefs and Practices Related to Disposal of Human Placenta. *Nurs J India* 80: 315–317.

Kambic, R. T., and V. M. Lamprecht. 1996. Calendar Rhythm Efficacy: A Review. *Adv Contraception* 12: 123–128.

Kang, J. H., F. Kondo, and Y. Katayama. 2006. Human Exposure to Bisphenol A. *Toxicology* 226: 79–89.

Katz, G. 1953. The Seasonal Variation in the Incidence of Premature Deliveries. *Nord Med* 50: 1638.

Katzenberg, M. A., D. A. Herring, and S. R. Saunders. 1996. Weaning and Infant Mortality: Evaluating the Skeletal Evidence. *Yearb Phys Anthropol* 39: 177–199.

Kenagy, G. J., and S. C. Trombulak. 1986. Size and Function of Mammalian Testes in Relation to Body Size. *J Mammal* 67: 1–22.

Kennedy, K. J., R. Rivera, and A. S. McNeilly. 1989. Consensus Statement on the Use of Breastfeeding as a Family Planning Method. *Contraception* 39: 477–496.

Kesserü, E. 1984. Sexual Intercourse Enhances the Success of Artificial Insemination. *Int J Fertil* 29: 143–145.

Khatamee, M. A. 1988. Infertility: A Preventable Epidemic. *Int J Fertil* 33: 246–251.

Kielan-Jaworowska, Z. 1979. Pelvic Structure and Nature of Reproduction in Multituberculata. *Nature* 277: 402–403.

Kiltie, R. A. 1982. Intraspecific Variation in the Mammalian Gestation Period. *J Mammal* 63: 646–652.

Kintner, H. J. 1985. Trends and Regional Differences in Breastfeeding in Germany from 1871–1937. *J Fam Med* 10: 163–182.

Klaus, M. H. 1987. The Frequency of Suckling: A Neglected but Essential Ingredient of Breast-feeding. *Obstet Gynecol Clin N Am* 14: 623–633.

Knodel, J. E. 1977. Breastfeeding and Population Growth: Assessing the Demographic Impact of Changing Infant Feeding Practices in the Third World. *Science* 198: 1111–1115.

Kobeissi, L., M. C. Inhorn, A. B. Hannoun, N. Hammoud, J. Awwad, and A. A. Abu-Musa. 2008. Civil War and Male Infertility in Lebanon. *Fertil Steril* 90: 340–345.

Koletzko, B., E. Lien, C. Agostoni, H. Böhles, C. I. Campoy, T. Decsi, J. W. Dudenhausen, C. Dupont, S. Forsyth, I. Hoesli, W. Holzgreve, A. Lapillonne, G. Putet, N. J. Secher, M. Symonds, H. Szajewska, P. Willatts, and R. Uauy. 2008. The Roles of Long-Chain Polyunsaturated Fatty Acids in Pregnancy, Lactation, and Infancy: Review of Current Knowledge and Consensus Recommendations. *J Perinat Med* 36: 5–14.

Konner, M., and C. Worthman. 1980. Nursing Frequency, Gonadal Function, and Birth Spacing Among !Kung Hunter-Gatherers. *Science* 207: 788–791.

Kovar, W. R., and R. J. Taylor. 1960. Is Spontaneous Abortion a Seasonal Problem? *Obst Gynecol* 16: 350–353.

Kramer, P. A. 1998. The Costs of Human Locomotion: Maternal Investment in Child Transport. *Am J Phys Anthropol* 107: 71–86.

Kunz, G., H. Deininger, L. Wildt, and G. Leyendecker. 1996. The Dynamics of Rapid Sperm Transport Through the Female Genital Tract: Evidence from Vaginal Sonography of Uterine Peristalsis and Hysterosalpingoscintigraphy. *Hum Reprod* 11: 627–632.

Kuruto-Niwa, R., Y. Tateoka, Y. Usuki, and R. Nozawa. 2007. Measurement of Bisphenol A Concentration in Human Colostrum. *Chemosphere* 66: 1160–1164.

Kuzawa, C. W. 1998. Adipose Tissue in Human Infancy and Childhood: An Evolutionary Perspective. *Yearb Phys Anthropol* 41: 177–209.

Labbok, M. H. 2001. Effects of Breastfeeding on the Mother [review]. *Pediatr Clin N Am* 48: 143–158.

Lam, D. A., and J. A. Miron. 1994. Global Patterns of Seasonal Variation in Human Fertility. *Ann NY Acad Sci* 709: 9–28.

Lansac, J., F. Thepot, M. J. Mayaux, F. Czyglick, T. Wack, J. Selva, and P. Jalbert. 1997. Pregnancy Outcome After Artificial Insemination or IVF With Frozen

Semen Donor: A Collaborative Study of the French CECOS Federation on 21,597 Pregnancies. *Eur J Obstet Gynecol Reprod Biol* 74: 223–228.

Lanting, C. I., V. Fidler, M. Huisman, B. C. L. Touwen, and E. R. Boersma. 1994. Neurological Differences Between 9-Year-Old Children Fed Breast-Milk or Formula-Milk as Babies. *Lancet* 344: 1319–1322.

Lau, C. 2001. Effects of Stress on Lactation. *Pediatr Clin N Am* 48: 221–234.

Lee, P. C. 1987. Nutrition, Fertility and Maternal Investment in Primates. *J Zool Lond* 213: 409–422.

Leigh, S. R., and P. B. Park. 1998. Evolution of Human Growth Prolongation. *Am J Phys Anthropol* 107: 331–350.

Lerchl, M., M. Simoni, and E. Nieschlag. 1993. Changes in the Seasonality of Birth Rates in Germany from 1951 to 1990. *Naturwiss* 80: 516–518.

Leutenegger, W. 1973. Maternal-Fetal Weight Relationships in Primates. *Folia Primatol* 20: 280–293.

Levin, R. J. 1975. Masturbation and Nocturnal Emissions: Possible Mechanisms for Minimising Teratospermie and Hyperspermie in Man. *Med Hypoth* 1: 130–131.

Lewy, A. J., T. A. Wehr, F. K. Goodwin, D. A. Newsome, and S. P. Markey. 1980. Light Suppresses Melatonin Secretion in Humans. *Science* 210: 1267–1269.

Li, D., Z. Zhou, D. Qing, Y. He, T. Wu, M. Miao, J. Wang, X. Weng, J. R. Ferber, L. J. Herrinton, Q. Zhu, E. Gao, H. Checkoway, and W. Yuan. 2010. Occupational Exposure to Bisphenol-A (BPA) and the Risk of Self-reported Male Sexual Dysfunction. *Hum Reprod* 25: 519–527.

Li, D.-K., Z.-J. Zhou, M. Miao, Y. He, J.-T. Wang, J. Ferber, L. J. Herrinton, E.-S. Gao, and W. Yuan. 2011. Urine Bisphenol-A (BPA) Level in Relation to Semen Quality. *Fertil Steril* 95: 625–630.

Lijeros, F., C. R. Edling, L. A. N. Amaral, H. E. Stanley, and Y. Åberg. 2001. The Web of Human Sexual Contacts. *Nature* 411: 907–908.

Linzenmeier, G. 1947. Zur Frage der Empfängniszeit der Frau: Hat Knaus oder Stieve recht? *Zentralbl Gynäkol* 69: 1108–1110.

Lloyd, J., N. S. Crouch, C. L. Minto, L.-M. Liao, and S. M. Creighton. 2005. Female Genital Appearance: "Normality" Unfolds. *Brit J Obstet Gynaecol* 112: 643–646.

Lönnerdal, B. 2000. Breast Milk: A Truly Functional Food. *Nutrition* 16: 509–511.

Lopata, A. 1996. Implantation of the Human Embryo. *Hum Reprod* 11 (Suppl. 1): 175–184.

Loucks, A. B., and L. M. Redman. 2004. The Effect of Stress on Menstrual Function. *Trends Endocrinol Metab* 15: 466–471.

Loudon, A. S. I., A. S. McNeilly, and J.A. Milne. 1983. Nutrition and Lactational Control of Fertility in Red Deer. *Nature* 302: 145–147.

Loy, J. 1987. The Sexual Behavior of African Monkeys and the Question of Estrus. In *Comparative Behavior of African Monkeys*, ed. E. Zucker, 175–195. New York: Alan Liss.

Luckett, W. P. 1974. The Comparative Development and Evolution of the Placenta in Primates. *Contrib Primatol* 3: 142–234.

MacDorman, M. F., F. Menacker, and E. Declercq. 2008. Cesarean Birth in the United States: Epidemiology, Trends, and Outcomes. *Clin Perinatol* 35: 293–307.

MacLeod, J., and R. Z. Gold. 1951. The Male Factor in Fertility and Infertility. II. Spermatozoon Counts in 1000 Men of Known Fertility and 1000 Cases of Infertile Marriage. *J Urol* 66: 436–449.

———. 1957. The Male Factor in Fertility and Infertility. IX. Semen Quality in Relation to Accidents of Pregnancy. *Fertil Steril* 8: 36–49.

MacLeod, J., and R. S. Hotchkiss. 1941. The Effect of Hyperpyrexia upon Spermatozoa Counts in Men. *Endocrinology* 28: 780–784.

MacLeod, J., and Y. Wang. 1979. Male Fertility Potential in Terms of Semen Quality. A Review of the Past, a Study of the Present. *Fertil Steril* 31: 103–116.

Macomber, D., and M. B. Sanders. 1929. The Spermatozoa Count. *N Engl J Med* 200: 981–984.

Málek, J., J. Gleich, and V. Maly. 1962. Characteristics of the Daily Rhythm of Menstruation and Labor. *Ann NY Acad Sci* 98: 1042–1055.

Mancuso, P. J., J. M. Alexander, D. D. McIntire, E. Davis, G. Burke, and K. J. Leveno. 2004. Timing of Birth After Spontaneous Onset of Labor. *Obstet Gynaecol* 103: 653–656.

Mann, D. R., and H. M. Fraser. 1996. The Neonatal Period: A Critical Interval in Male Primate Development. *J Endocrinol* 149: 191–197.

Marshall, J. 1968. Congenital Defects and the Age of Spermatozoa. *Int J Fertil* 13: 110–120.

Martin, R. D. 1968. Reproduction and Ontogeny in Tree-Shrews (*Tupaia belangeri*) with Reference to Their General Behaviour and Taxonomic Relationships. *Z Tierpsychol* 25: 409–532.

———. 1969. The Evolution of Reproductive Mechanisms in Primates. *J Reprod Fertil Suppl* 6: 49–66.

———. 1981. Relative Brain Size and Metabolic Rate in Terrestrial Vertebrates. *Nature* 293: 57–60.

———. 1984. Scaling Effects and Adaptive Strategies in Mammalian Lactation. *Symp Zool Soc Lond* 51: 87–117.

———. 1992. Female Cycles in Relation to Paternity in Primate Societies. In *Paternity in Primates: Genetic Tests and Theories. Implications of Human DNA Fingerprinting*, ed. R. D. Martin, A. F. Dixson, and E. J. Wickings, 238–274. Basel: Karger.

———. 1996. Scaling of the Mammalian Brain: The Maternal Energy Hypothesis. *News Physiol Sci* 11: 149–156.

———. 2003. Human Reproduction: A Comparative Background for Medical Hypotheses. *J Reprod Immunol* 59: 111–135.

———. 2007. The Evolution of Human Reproduction: A Primatological Perspective. *Yearb Phys Anthropol* 50: 59–84.

———. 2008. Evolution of Placentation in Primates: Implications of Mammalian Phylogeny. *Evol Biol* 35: 125–145.

———. 2012. Primer: Primates. *Curr Biol* 22: R785–R790.

Martin, R. D., and K. Isler. 2010. The Maternal Energy Hypothesis of Brain Evolution: An Update. In *The Human Brain Evolving: Paleoneurological Studies in Honor of Ralph L. Holloway*, ed. D. Broadfield, M. Yuan, K. Schick, and N. Toth, 15–35. Bloomington, IN: Stone Age Institute Press.

Martin, R. D., and A. M. MacLarnon. 1985. Gestation Period, Neonatal Size and Maternal Investment in Placental Mammals. *Nature* 313: 220–223.

———. 1988. Comparative Quantitative Studies of Growth and Reproduction. *Symp Zool Soc Lond* 60: 39–80.

Martin, R. D., L. A. Willner, and A. Dettling. 1994. The Evolution of Sexual Size Dimorphism in Primates. In *The Differences Between the Sexes*, ed. R. V. Short and E. Balaban, 159–200. Cambridge: Cambridge University Press.

Matsuda, S., and H. Kahyo. 1992. Seasonality of Preterm Births in Japan. *Int J Epidemiol* 21: 91–100.

McCance, R. A., M. C. Luff, and E. E. Widdowson. 1937. Physical and Emotional Periodicity in Women. *J Hygiene* 37: 571–611.

McCoy, S. J. B., J. M. Beal, S. B. M. Shipman, M. E. Payton, and G. H. Watson. 2006. Risk Factors for Postpartum Depression: A Retrospective Investigation at 4-Weeks Postnatal and a Review of the Literature. *J Am Osteopath Assoc* 106: 193–198.

McGrath, J. J., A. G. Barnett, and D. W. Eyles. 2005. The Association Between Birth Weight, Season of Birth and Latitude. *Ann Hum Biol* 32: 547–559.

McKenna, J. J., H. L. Ball, and L. T. Gettler. 2007. Mother-Infant Cosleeping, Breastfeeding and Sudden Infant Death Syndrome: What Biological Anthropology Has Discovered About Normal Infant Sleep and Pediatric Sleep Medicine. *Yearb Phys Anthropol* 45: 133–161.

McKeown, T., and J. R. Gibson. 1951. Observations on All Births (23,970) in Birmingham, 1947. II: Birth Weight. *Brit J Soc Med* 5: 98–112.

———. 1952. Period of Gestation. *Brit Med J* 1: 938–941.

McKeown, T., and R. G. Record. 1952. Observations on Foetal Growth in Multiple Pregnancy in Man. *J Endocrinol* 8: 386–401.

McNeilly, A. S. 2001. Lactational Control of Reproduction. *Reprod Fert Dev* 13: 583–590.

McTiernan, A., and D. B. Thomas. 1986. Evidence for a Protective Effect of Lactation on Risk of Breast Cancer in Young Women. Results from a Case Control Study. *Am J Epidemiol* 124: 353–358.

Menacker, F., and B. E. Hamilton. 2010. Recent Trends in Cesarean Delivery in the United States. *NCHS Data Brief* 35: 1–8.

Mendiola, J., N. Jørgensen, A.-M. Andersson, A. M. Calafat, X. Ye, J. B. Redmon, E. Z. Drobnis, C. Wang, A. Sparks, S. W. Thurston, and S. H. Swan. 2010. Are Environmental Levels of Bisphenol A Associated with Reproductive Function in Fertile Men? *Environm Health Perspect* 118: 1286–1291.

Michael, R. P., and E. B. Keverne. 1971. An Annual Rhythm in the Sexual Activity of the Male Rhesus Monkey, *Macaca mulatta*, in the Laboratory. *J Reprod Fertil* 25: 95–98.

Michaelsen, K. F., L. Lauritzen, and E. L. Mortensen. 2009. Effects of Breastfeeding on Cognitive Function. *Adv Exp Med Biol* 639: 199–215.

Mieusset, R., and L. Bujan. 1994. The Potential of Mild Testicular Heating as a Safe, Effective and Reversible Contraceptive Method for Men. *Int J Androl* 17: 186–191.

Miller, J. F., E. Williamson, J. Glue, Y. B. Gordon, J. G. Grudzinskas, and A. Sykes. 1980. Fetal Loss After Implantation. *Lancet* 316: 554–556.

Milligan, L. A., and R. P. Bazinet. 2008. Evolutionary Modifications of Human Milk Composition: Evidence from Long-Chain Polyunsaturated Fatty Acid Composition of Anthropoid Milks. *J Hum Evol* 55: 1086–1095.

Milstein-Moscati, I., and W. Beçak. 1978. Down Syndrome and Frequency of Intercourse. *Lancet* 312: 629–630.

———. 1981. Occurrence of Down Syndrome and Human Sexual Behavior. *Am J Med Genet* 9: 211–217.

Moffett, A., and Y. W. Loke. 2006. Immunology of Placentation in Eutherian Mammals. *Nature Rev Immunol* 6: 584–594.

Moghissi, K. S. 1976. Accuracy of Basal Body Temperature for Ovulation Detection. *Fertil Steril* 27: 1415–1421.

Møller, A. P. 1988. Ejaculate Quality, Testes Size and Sperm Competition in Primates. *J Hum Evol* 17: 479–488.

———. 1989. Ejaculate Quality, Testes Size and Sperm Production in Mammals. *Funct Ecol* 3: 91–96.

Montagu, A. 1961. Neonatal and Infant Immaturity in Man. *JAMA* 178: 56.

Morrow-Tlucak, M., R. H. Haude, and C. B. Ernhart. 1988. Breastfeeding and Cognitive Development in the First 2 Years of Life. *Soc Sci Med* 26: 635–639.

Mortensen, E. L., K. F. Michaelson, S. A. Sanders, and J. M. Reinisch. 2002. The Association Between Duration of Breastfeeding and Adult Intelligence. *J Am Med Ass* 287: 2365–2371.

Mortimer, D. 1983. Sperm Transport in the Human Female Reproductive Tract. *Oxford Rev Reprod Biol* 5: 30–61.

Mulcahy, M. T. 1978. Down Syndrome and Parental Coital Rate. *Lancet* 312: 895.

Munshi-South, J. 2007. Extra-Pair Paternity and the Evolution of Testis Size in a Behaviorally Monogamous Tropical Mammal, the Large Treeshrew (*Tupaia tana*). *Behav Ecol Sociobiol* 62: 201–212.

Münster, K., L. Schmidt, and P. Helm. 1992. Length and Variation in the Menstrual Cycle—A Cross-Sectional Study from a Danish County. *Brit J Obstet Gynecol* 99: 422–429.

Nadler, R. D. 1994. Walter Heape and the Issue of Estrus in Primates. *Am J Primatol* 33: 83–99.

Nathanielsz, P. W. 1998. Comparative Studies on the Initiation of Labor. *Europ J Obstet Gynecol Reprod Biol* 78: 127–132.

Nelson, C. M. K., and R. G. Bunge. 1974. Semen Analysis: Evidence for Changing Parameters of Male Fertility Potential. *Fertil Steril* 25: 503–507.

Neugebauer, F. L. 1886. Eine bisher einzig dastehende Beobachtung von Polymastie mit 10 Brustwarzen. *Zentralbl Gynäkol* 10: 720–736.

Newman, J. 1995. How Breast Milk Protects Newborns. *Sci Am* 273(12): 76–79.

Oatridge, A., A. Holdcroft, N. Saeed, J. V. Hajnal, B. K. Puri, L. Fusi, and G. M. Bydder. 2002. Change in Brain Size During and After Pregnancy: Study in Healthy Women and Women with Preeclampsia. *Am J Neuroradiol* 23: 19–26.

Odeblad, E. 1997. Cervical Mucus and Their Functions. *J Irish Coll Phys Surg* 26: 27–32.

Oftedal, O. T., and S. J. Iverson. 1995. Phylogenetic Variation in the Gross Composition of Milks. In *The Handbook of Milk Composition*, ed. R. G. Jensen, M. P. Thompson, and R. Jenness, 749–789. Orlando, FL: Academic Press.

O'Hara, M. W., and A. M. Swain. 1996. Rates and Risk of Postpartum Depression—A Metaanalysis. *Int Rev Psychiatr* 8: 37–54.

O'Rand, M. G., E. E. Widgren, S. Beyler, and R. T. Richardson. 2009. Inhibition of Human Sperm Motility by Contraceptive Anti-eppin Antibodies from Infertile Male Monkeys: Effect on Cyclic Adenosine Monophosphate. *Biol Reprod* 80: 279–285.

Papanicolaou, G. N. 1933. The Sexual Cycle of the Human Female as Revealed by Vaginal Smears. *Am J Anat* 52: 519–637.

Paraskevaides, E. C., G. W. Pennington, and S. Naik. 1988. Seasonal Distribution in Conceptions Achieved by Artificial Insemination by Donor. *Brit Med J* 297: 1309–1310.

Parazzini, F., M. Marchini, L. Luchini, L. Tozzi, R. Mezzopane, and L. Fedele. 1995. Tight Underpants and Trousers and the Risk of Dyspermia. *Int J Androl* 18: 137–140.

Parente, R. C. M., L. P. Bergqvist, M. B. Soares, and O. B. Moraes. 2011. The History of Vaginal Birth. *Arch Gynecol Obstet* 284: 1–11.

Parker, G. A. 1982. Why So Many Tiny Sperm? The Maintenance of Two Sexes with Internal Fertilization. *J Theor Biol* 96: 281–294.

Pawłowski, B. 1998. Why Are Human Newborns So Big and Fat? *Hum Evol* 13: 65–72.

———. 1999. Permanent Breasts as a Side Effect of Subcutaneous Fat Tissue Increase in Human Evolution. *Homo* 50: 149–162.

Pearson, J. A., R. E. M. Hedges, T. I. Molleson, and M. Özbek. 2010. Exploring the Relationship Between Weaning and Infant Mortality: An Isotope Case Study from Asikli Höyük and Cayönü Tepesi. *Am J Phys Anthropol* 143: 448–457.

Penrose, L. S., and J. M. Berg. 1968. Mongolism and Duration of Marriage. *Nature* 218: 300.

Pepper, G. V., and S. C. Roberts. 2006. Rates of Nausea and Vomiting in Pregnancy and Dietary Characteristics Across Populations. *Proc Roy Soc Lond B* 273: 2675–2679.

Piovanetti, Y. 2001. Breastfeeding Beyond 12 Months: An Historical Perspective. *Pediatr Clin N Am* 48: 199–206.

Plavcan, J. M. 2012. Sexual Size Dimorphism, Canine Dimorphism, and Male-Male Competition in Primates: Where Do Humans Fit In? *Hum Nat* 23: 45–67.

Poikkeus, P., M. Gissler, L. Unkila-Kallio, C. Hyden-Granskog, and A. Tiitinen. 2007. Obstetric and Neonatal Outcome After Single Embryo Transfer. *Hum Reprod* 22: 1073–1079.

Ponce de León, M. S., L. Golovanova, V. Doronichev, G. Romanova, T. Akazawa, O. Kondo, H. Ishida, and C. P. E. Zollikofer. 2008. Neanderthal Brain Size at Birth Provides Insights into the Evolution of Human Life History. *Proc Natl Acad Sci USA* 105: 13764–13768.

Procopé, B.-J. 1965. Effect of Repeated Increase of Body Temperature on Human Sperm Cells. *Int J Fertil* 10: 333–339.

Profet, M. 1993. Menstruation as a Defence Against Pathogens Transported by Sperm. *Quart Rev Biol* 68: 335–386.

Racey, P. A. 1979. The Prolonged Storage and Survival of Spermatozoa in Chiroptera. *J Reprod Fertil* 56: 391–402.

Ramlau-Hansen, C. H., G. Toft, M. S. Jensen, K. Strandberg-Larsen, M. L. Hansen, and J. Olsen. 2010. Maternal Alcohol Consumption During Pregnancy and Semen Quality in the Male Offspring: Two Decades of Follow-Up. *Hum Reprod* 25: 2340–2345.

Ramm, S. A. 2007. Sexual Selection and Genital Evolution in Mammals: A Phylogenetic Analysis of Baculum Length. *Am Nat* 169: 360–369.

Ramm, S. A., P. L. Oliver, C. P. Ponting, P. Stockley, and R. D. Emes. 2008. Sexual Selection and the Adaptive Evolution of Mammalian Ejaculate Proteins. *Mol Biol Evol* 25: 207–219.

Record, R. G. 1952. Relative Frequencies and Sex Distributions of Human Multiple Births. *Brit J Soc Med* 6: 192–196.

Reefhuis, J., M. A. Honein, L. A. Schieve, A. Correa, C. A. Hobbs, S. A. Rasmussen, and National Birth Defects Prevention Study. 2009. Assisted Reproductive Technology and Major Structural Birth Defects in the United States. *Hum Reprod* 24: 360–366.

Rehan, N., A. J. Sobbero, and J. W. Fertig. 1975. The Semen of Fertile Men: Statistical Analysis of 1300 Men. *Fertil Steril* 26: 492–502.

Reinberg, A. 1974. Aspects of Circannual Rhythms in Man. In *Circannual Clocks: Annual Biological Rhythms*, ed. E. T. Pengelley, 423–505. New York: Academic Press.

Reinberg, A., and M. Lagoguey. 1978. Circadian and Circannual Rhythms in Sexual Activity and Plasma Hormones (FSH-LH, Testosterone) of Five Human Males. *Arch Sex Behav* 7: 13–30.

Renaud, R. L., J. Macler, I. Dervain, M.-C. Ehret, C. Aron, S. Plas-Roser, A. Spira, and H. Pollack. 1980. Echograpic Study of Follicular Maturation and Ovulation During the Normal Menstrual Cycle. *Fertil Steril* 33: 272–276.

Reynolds, A. 2001. Breastfeeding and Brain Development. *Pediatr Clin N Am* 48: 159–171.

Richard, A. F. 1974. Patterns of Mating in *Propithecus verreauxi*. In *Prosimian Biology*, ed. R. D. Martin, G. A. Doyle, and A. C. Walker, 49–75. London: Duckworth.

Riggs, R., J. Mayer, D. Dowling-Lacey, T.-F. Chi, E. Jones, and S. Oehninger. 2010. Does Storage Time Influence Postthaw Survival and Pregnancy Outcome? An Analysis of 11,768 Cryopreserved Human Embryos. *Fertil Steril* 93: 109–115.

Roberts, C., and C. Lowe. 1975. Where Have All the Conceptions Gone? *Lancet* 305: 498–499.

Robinson, D., J. Rock, and M. F. Menkin. 1968. Control of Human Spermatogenesis by Induced Changes in Intrascrotal Temperature. *JAMA* 204: 290–297.

Rock, J. C., and D. Robinson. 1965. Effect of Induced Intrascrotal Hyperthermia on Testicular Function in Man. *Am J Obstet Gynecol* 93: 793–801.

Rodgers, B. 1978. Feeding in Infancy and Later Ability and Attainment: A Longitudinal Study. *Dev Med Child Neurol* 20: 421–426.

Roenneberg, T., and J. Aschoff. 1990a. Annual Rhythm of Human Reproduction. I. Biology, Sociobiology or Both? *J Biol Rhythms* 5: 195–216.

———. 1990b. Annual Rhythm of Human Reproduction. II. Environmental Correlations. *J Biol Rhythms* 5: 217–239.

Rogan, J. W., and B. C. Gladen. 1993. Breast Feeding and Cognitive Development. *Early Hum Dev* 31: 181–193.

Rojansky, N., A. Brzezinski, and J. G. Schenker. 1992. Seasonality in Human Reproduction: An Update. *Hum Reprod* 7: 735–745.

Rolland, M., J. Moal, V. Wagner, D. Royère, and J. De Mouzon. 2012. Decline in Semen Concentration and Morphology in a Sample of 26 609 Men Close to General Population between 1989 and 2005 in France. *Hum Reprod* 28: 462–470.

Ron-El, R., A. Golan, H. Nachum, E. Caspi, A. Herman, and Y. Softer. 1991. Delayed Fertilization and Poor Embryonic Development Associated with Impaired Semen Quality. *Fertil Steril* 55: 338–344.

Rosenberg, K. R. 1992. The Evolution of Modern Human Childbirth. *Yearb Phys Anthropol* 35: 89–124.

Rosenberg, K. R., and W. Trevathan. 1996. Bipedalism and Human Birth: The Obstetrical Dilemma Revisited. *Evol Anthropol* 4: 161–168.

———. 2002. Birth, Obstetrics and Human Evolution. *Brit J Obstet Gynaecol* 109: 1199–1206.

Rowley, M. J., F. Teshima, and C. G. Heller. 1970. Duration of Transit of Spermatozoa Through the Human Male Ductular System. *Fertil Steril* 21: 390–396.

Rubenstein, B. B., H. Strauss, M. L. Lazarus, and H. Hankin. 1951. Sperm Survival in Women: Motile Sperm in the Fundus and the Tubes of Surgical Cases. *Fertil Steril* 2: 15–19.

Sacher, G. A. 1982. The Role of Brain Maturation in the Evolution of the Primates. In *Primate Brain Evolution*, ed. E. Armstrong and D. Falk, 97–112. New York: Plenum.

Sacher, G. A., and E. F. Staffeldt. 1974. Relation of Gestation Time to Brain Weight for Placental Mammals: Implications for the Theory of Vertebrate Growth. *Am Nat* 108: 593–615.

Sade, D. S. 1964. Seasonal Cycle in Size of Testes of Free-Ranging *Macaca mulatta*. *Folia Primatol* 2: 171–180.

Sanders, D., and J. Bancroft. 1982. Hormones and the Sexuality of Women—The Menstrual Cycle. *Clin Endocrinol Metab* 11: 639–659.

Sas, M., and J. Szöllősi. 1979. Impaired Spermiogenesis as a Common Finding Among Professional Drivers. *Arch Androl* 3: 57–60.

Schaffir, J. 2006. Sexual Intercourse at Term and Onset of Labor. *Obstet Gynaecol* 107: 1310–1314.

Schernhammer, E. S., and S. E. Hankinson. 2005. Urinary Melatonin Levels and Breast Cancer Risk. *J Nat Cancer Inst* 97: 1084–1087.

Schiebinger, L. 1993. Why Mammals Are Called Mammals: Gender Politics in Eighteenth-Century Natural History. *Am Hist Rev* 90: 382–411.

Schneiderman, J. U. 1998. Rituals of Placenta Disposal. *Am J Matern Child Nurs* 23: 142–143.

Schradin, C., and G. Anzenberger. 2001. Costs of Infant Carrying in Common Marmosets, *Callithrix jacchus*: An Experimental Analysis. *Anim Behav* 62: 289–295.

Sellen, D. W. 2001. Comparison of Infant Feeding Patterns Reported for Nonindustrial Populations with Current Recommendations. *J Nutr* 131: 2707–2715.

———. 2009. Evolution of Human Lactation and Complementary Feeding: Implications for Understanding Contemporary Cross-Cultural Variation. *Adv Exp Med Biol* 639: 253–282.

Setchell, B. P. 1997. Sperm Counts in Semen of Farm Animals 1932–1995. *Int J Androl* 20: 209–214.

———. 1998. The Parkes Lecture: Heat and the Testes. *J Reprod Fertil* 114: 179–194.

Settlage, D. S. F., M. Motoshima, and D. R. Tredway. 1973. Sperm Transport from the External Cervical Os to the Fallopian Tubes in Women. *Fertil Steril* 24: 655–661.

Shafik, A. 1992. Contraceptive Efficacy of Polyester-Induced Azoospermia in Normal Men. *Contraception* 45: 439–451.

Sharav, T. 1991. Aging Gametes in Relation to Incidence, Gender, and Twinning in Down Syndrome. *Am J Med Genet* 39: 116–118.

Sharpe, R. M. 1994. Could Environmental, Oestrogenic Chemicals Be Responsible for Some Disorders of Human Male Reproductive Development? *Curr Opin Urol* 4: 295–302.

Sharpe, R. M., and N. E. Skakkebaek. 1993. Are Oestrogens Involved in Falling Sperm Counts and Disorders of the Male Reproductive Tract? *Lancet* 341: 1392–1395.

Sheard, N. F., and W. A. Walker. 1988. The Role of Breast Milk in the Development of the Gastrointestinal Tract. *Nutr Rev* 46: 1–8.

Sheynkin, Y., R. Welliver, A. Winer, F. Hajimirzaee, H. Ahn, and K. Lee. 2011. Protection from Scrotal Hyperthermia in Laptop Computer Users. *Fertil Steril* 95: 647–651.

Short, R. V. 1976. The Evolution of Human Reproduction. *Proc R Soc Lond B Biol Sci* 195: 3–24.

———. 1979. Sexual Selection and Its Component Parts, Somatic and Genital Selection, as Illustrated by Man and the Great Apes. *Adv Stud Behav* 9: 131–158.

———. 1984. Breast Feeding. *Sci Am* 250, 4: 35–41.

———. 1994. Human Reproduction in an Evolutionary Context. *Ann NY Acad Sci* 709: 416–425.

Simmons, L. W., L. C. Firman, G. Rhodes, and M. Peters. 2004. Human Sperm Competition: Testis Size, Sperm Production and Rates of Extrapair Copulations. *Anim Behav* 68: 297–302.

Simpson, J. L., R. H. Gray, A. Perez, P. Mena, M. Barbato, E. E. Castilla, R. T. Kambic, F. Pardo, G. Tagliabue, W. S. Stephenson, A. Bitto, C. Li, V. H. Jennings, J. M. Spieler, and J. T. Queenan. 1997. Pregnancy Outcome in Natural Family Planning Users: Cohort and Case-Control Studies Evaluating Safety. *Adv Contraception* 13: 201–214.

Slama, R., F. Eustache, B. Ducot, T. K. Jensen, N. Jørgensen, A. Horte, S. Irvine, J. Suominen, A. G. Andersen, J. Auger, M. Vierula, J. Toppari, J. N. Andersen, N. Keiding, N. E. Skakkebaek, A. Spira, and P. Jouannet. 2002. Time to Pregnancy and Semen Parameters: A Cross-Sectional Study Among Fertile Couples from Four European Cities. *Hum Reprod* 17: 503–515.

Small, M. 1992. The Evolution of Female Sexuality and Mate Selection in Humans. *Hum Nat* 3: 133–156.

———. 1996. "Revealed" Ovulation in Humans? *J Hum Evol* 30: 483–488.

Smits, J., and C. Monden. 2011. Twinning Across the Developing World. *PLoS One* 6(9): e25239.

Sokol, R. Z., P. Kraft, I. M. Fowler, R. Mamet, E. Kim, and K. T. Berhane. 2006. Exposure to Environmental Ozone Alters Semen Quality. *Environ Health Perspect* 114: 360–365.

Spira, A. 1984. Seasonal Variations of Sperm Characteristics. *Arch Androl* 12 (Suppl): 23–28.

Stallmann, R. R., and A. H. Harcourt. 2006. Size Matters: The (Negative) Allometry of Copulatory Duration in Mammals. *Biol J Linn Soc* 87: 185–193.

Stanislaw, H., and F. J. Rice. 1988. Correlation Between Sexual Desire and Menstrual Cycle Characteristics. *Arch Sex Behav* 17: 499–508.

Steklis, H. D., and C. H. Whiteman. 1989. Loss of Estrus in Human Evolution: Too Many Answers, Too Few Questions. *Ethol Sociobiol* 10: 417–434.

Stephens, W. N. 1961. A Cross-Cultural Study of Menstrual Taboos. *Genet Psychol Monogr* 64: 385–416.

Steptoe, P. C., and R. G. Edwards. 1978. Birth After the Reimplantation of a Human Embryo. *Lancet* 312: 366.

Storgaard, L., J. Bonde, E. Ernst, M. Spano, C. Y. Andersen, M. Frydenberg, and J. Olsen. 2003. Does Smoking During Pregnancy Affect Sons' Sperm Counts? *Epidemiology* 14: 278–286.

Strassmann, B. I. 1996a. Energy Economy in the Evolution of Menstruation. *Evol Anthropol* 5: 157–164.

———. 1996b. The Evolution of Endometrial Cycles and Menstruation. *Quart Rev Biol* 71: 181–220.

———. 1997. The Biology of Menstruation in *Homo sapiens*: Total Lifetime Menses, Fecundity, and Nonsynchrony in a Natural-Fertility Population. *Curr Anthropol* 38: 123–129.

Suarez, S. S., and A. A. Pacey. 2006. Sperm Transport in the Female Reproductive Tract. *Hum Reprod Update* 12: 23–37.

Sugarman, M., and K. A. Kendall-Tackett. 1995. Weaning Ages in a Sample of American Women Who Practice Extended Breastfeeding. *Clin Pediatr (Philadelphia)* 34: 642–647.

Suomi, S. J., and C. Ripp. 1983. A History of Motherless Mothering at the University of Wisconsin Primate Laboratory. In *Child Abuse: The Nonhuman Primate Data*, ed. M. Reite and N. G. Caine, 49–78. New York: Alan Liss.

Swan, S. H., E. P. Elkin, and L. Fenster. 2000. The Question of Declining Sperm Density Revisited: An Analysis of 101 Studies Published 1934–1996. *Environm Health Persp* 108: 961–966.

Sydenham, A. 1946. Amenorrhoea at Stanley Camp, Hong Kong, During Internment. *Brit Med J* 2: 159.

Thiery, M. 2000. Intrauterine Contraception: From Silver Ring to Intrauterine Contraceptive Implant. *Eur J Obstet Gynecol Reprod Biol* 90: 145–152.

Tiemessen, C. H. J., J. L. Evers, and R. S. G. M. Bots. 1995. Tight Fitting Underwear and Sperm Quality. *Lancet* 347: 1844–1845.

Tietze, C. 1965. History of Contraceptive Methods. *J Sex Res* 1: 69–85.

Topinard, P. 1882a. Le Poids du Cerveau d'après les Registres de Paul Broca. *Rev d'Anthropol, sér 2* 5: 1–30.

———. 1882b. La Mensuration de la Capacité du Crâne. *Rev d'Anthropol, sér 2* 5: 385–411.

Treloar, A. E., R. E. Boynton, B. G. Behn, and B. W. Brown. 1967. Variation of the Human Menstrual Cycle Through Reproductive Life. *Int J Fertil* 12: 77–126.

Trevathan, W. R. 2007. Evolutionary Medicine. *Ann Rev Anthropol* 36: 139–154.

Trinkaus, E. 1984. Neandertal Public Morphology and Gestation Length. *Curr Anthropol* 25: 509–513.

Trussell, J. 2011. Contraceptive Failure in the United States. *Contraception* 83: 397–404.

Trussell, J., R. A. Hatcher, W. Cates, F. H. Stewart, and K. Kost. 1990. Contraceptive Failure in the United States: An Update. *Stud Fam Plann* 21: 51–54.

Tummon, I. S., and D. Mortimer. 1992. Decreasing Quality of Semen. *Brit Med J* 305: 1228–1229.

Tycko, B., and A. Efstratiadis. 2002. Genomic Imprinting: Piece of Cake. *Nature* 417: 913–914.

Tyler, E. T. 1953. Physiological and Clinical Aspects of Conception. *J Am Med Ass* 153: 1351–1356.

Udry, J. R., and N. M. Morris. 1967. Seasonality of Coitus and Seasonality of Birth. *Demography* 4: 673–679.

———. 1968. Distribution of Coitus in the Menstrual Cycle. *Nature* 220: 593–596.

———. 1977. The Distribution of Events in the Human Menstrual Cycle. *J Reprod Fertil* 51: 419–425.

Vandenberg, L. N., I. Chahoud, J. J. Heindel, V. Padmanabhan, F. J. R. Paumgartten, and G. Schoenfelder. 2010. Urinary, Circulating, and Tissue Biomonitoring Studies Indicate Widespread Exposure to Bisphenol A. *Environm Health Perspect* 118: 1055–1070.

van Os, J. L., M. J. de Vries, N. H. den Daas, and L. M. K. Lansbergen. 1997. Longterm Trends in Sperm Counts in Dairy Bulls. *J Androl* 18: 725–731.

Vennemann, M. M., T. Bajanowski, B. Brinkmann, G. Jorch, K. Yücesan, C. Sauerland, E. A. Mitchell, and the GeSID Study Group. 2009. Does Breastfeeding Reduce the Risk of Sudden Infant Death Syndrome? *Pediatrics* 123: e406–e410.

Viterbo, P. 2004. I Got Rhythm: Gershwin and Birth Control in the 1930s. *Endeavour* 28: 30–35.

Vitzthum, V. J. 1994. Comparative Study of Breastfeeding Structure and Its Relation to Human Reproductive Ecology. *Yearb Phys Anthropol* 37: 307–349.

von Holst, D. 1974. Social Stress in the Tree-Shrew: Its Causes and Physiological and Ethological Consequences. In *Prosimian Biology*, ed. R. D. Martin, G. A. Doyle, and A. C. Walker, 389–411. London: Duckworth.

Waldinger, M. D., P. Quinn, M. Dilleen, R. Mundayat, D. H. Schweitzer, and M. Boolell. 2005. A Multinational Population Survey of Intravaginal Ejaculation Latency Time. *J Sex Med* 2: 492–497.

Wang, Y. S., and S. Y. Wu. 1996. The Effect of Exclusive Breastfeeding on Development and Incidence of Infection in Infants. *J Hum Lact* 12: 27–30.

Weaver, T. D., and J.-J. Hublin. 2009. Neandertal Birth Canal Shape and the Evolution of Human Childbirth. *Proc Natl Acad Sci USA* 106: 8151–8156.

Wehr, T. A. 1991. The Durations of Human Melatonin Secretion and Sleep Respond to Changes in Daylength (Photoperiod). *J Clin Endocrinol Metab* 73: 1276–1280.

———. 2001. Photoperiodism in Humans and Other Primates: Evidence and Implications. *J Biol Rhythms* 16: 348–364.

Wehr, T. A., H. A. Giesen, D. E. Moul, E. H. Turner, and P. J. Schwartz. 1995. Suppression of Men's Responses to Seasonal Changes in Day-Length by Modern Artificial Lighting. *Am J Physiol* 269: R173–R178.

Weigel, R. M., and M. M. Weigel. 1989. Nausea and Vomiting of Early Pregnancy and Pregnancy Outcome: A Meta-analytical Review. *Brit J Obstet Gynaecol* 96: 1312–1318.

Weiss, K. 2004. The Frog in Taffeta Pants. *Evol Anthropol* 13: 5–10.

Westoff, C. F. 1976. The Decline of Unplanned Births in the United States. *Science* 91: 38–41.

Whitcome, K., and D. E. Lieberman. 2007. Fetal Load and the Evolution of Lumbar Lordosis in Bipedal Hominins. *Nature* 450: 1075–1078.

White, D. R., E. M. Widdowson, H. Q. Woodard, and J. W. T. Dickerson. 1991. The Composition of Body Tissues (II). Fetus to Young Adult. *Brit J Radiol* 64: 149–159.

Wickings, E. J., and E. Nieschlag. 1980. Seasonality in Endocrine and Exocrine Function of the Adult Rhesus Monkey (*Macaca mulatta*) Maintained in a Controlled Laboratory Environment. *Int J Androl* 3: 87–104.

Wilcox, A. J., D. D. Baird, D. B. Dunson, D. R. McConnaughey, J. S. Kesner, and C. R. Weinberg. 2004. On the Frequency of Intercourse Around Ovulation: Evidence for Biological Influences. *Hum Reprod* 19: 1539–1543.

Wilcox, A. J., D. Dunson, and D. D. Baird. 2000. The Timing of the "Fertile Window" in the Menstrual Cycle: Day Specific Estimates from a Prospective Study. *Brit Med J* 321: 1259–1262.

Wilcox, A. J., C. R. Weinberg, J. F. O'Connor, D. D. Baird, J. P. Schlatterer, R. E. Canfield, E. G. Armstrong, and B. C. Nisula. 1988. Incidence of Early Loss of Pregnancy. *New Engl J Med* 319: 189–194.

Williams, G. C., and R. M. Nesse. 1991. The Dawn of Darwinian Medicine. *Quart Rev Biol* 66: 1–22.

Williams, M., C. J. Hill, I. Scudamore, B. Dunphy, I. D. Cooke, and C. L. R. Barratt. 1993. Sperm Numbers and Distribution Within the Human Fallopian Tube Around Ovulation. *Hum Reprod* 8: 2019–2026.

Wittmann, M., J. Dinich, M. Merrow, and T. Roenneberg. 2006. Social Jetlag: Misalignment of Biological and Social Time. *Chronobiol Int* 23: 497–509.

Wolf, D. P., W. Byrd, P. Dandekar, and M. M. Quigley. 1984. Sperm Concentration and the Fertilization of Human Eggs In Vitro. *Biol Reprod* 31: 837–848.

Wolff, P. H. 1968a. The Serial Organization of Sucking in the Young Infant. *Pediatrics* 42: 943–956.

———. 1968b. Sucking Patterns of Infant Mammals. *Brain Behav Evol* 1: 354–367.

Wood, J. W. 1989. Sperm Longevity. *Oxf Rev Reprod Biol* 11: 61–109.

Wood, S., A. Quinn, S. Troupe, C. Kingsland, and I. Lewis-Jones. 2006. Seasonal Variation in Assisted Conception Cycles and the Influence of Photoperiodism on Outcome in In Vitro Fertilization Cycles. *Hum Fertil* 9: 223–229.

Work Group on Breastfeeding. 1997. Breastfeeding and the Use of Human Milk. *Pediatrics* 100: 1035–1039.

World Health Organisation. 1985. Appropriate Technology for Birth. *Lancet* 326: 436–437.

Wright, L. E., and H. P. Schwarcz. 1998. Stable Carbon and Oxygen Isotopes in Human Tooth Enamel: Identifying Breastfeeding and Weaning in Prehistory. *Am J Phys Anthropol* 106: 1–18.

Yang, C. P. 1993. History of Lactation and Breast Cancer Risk. *Am J Epidemiol* 138: 1050–1056.

Yoshida, Y. 1960. Studies on Single Insemination with Donor's Semen. *J Jap Obstet Gynecol Soc* 7: 19–34.

Young, S. M., D. C. Benyshek, and P. Lienard. 2012. The Conspicuous Absence of Placenta Consumption in Human Postpartum Females: The Fire Hypothesis. *Ecol Food Nutr* 51: 198–217.

Zalko, D., C. Jacques, H. Duplan, S. Bruel, and P. Perdu. 2011. Viable Skin Efficiently Absorbs and Metabolizes Bisphenol A. *Chemosphere* 82: 424–430.

Zeilmaker, G. H., A. T. Alberda, I. Vangent, C. M. P. M. Rijkmans, and A. C. Drogendijk. 1984. 2 Pregnancies Following Transfer of Intact Frozen-Thawed Embryos. *Fertil Steril* 42: 293–296.

Zhang, X., C. Zhu, H. Lin, Q. Yang, Q. Ou, Y. Li, Z. Chen, P. Racey, S. Zhang, and H. Wang. 2007. Wild Fulvous Fruit Bats (*Rousettus leschenaulti*) Exhibit Human-like Menstrual Cycle. *Biol Reprod* 77: 358–364.

Ziegler, E. E., A. M. O'Donnell, S. E. Nelson, and S. J. Fomon. 1976. Body Composition of the Reference Fetus. *Growth* 40: 329–341.

Zimmer, C. 2009. On the Origin of Sexual Reproduction. *Science* 324: 1254–1256.

Zinaman, M., E. Z. Drobnis, P. Morales, C. Brazil, M. Kiel, N. L. Cross, F. W. Hanson, and J. W. Overstreet. 1989. The Physiology of Sperm Recovered from the Human Cervix: Acrosomal Status and Response to Inducers of the Acrosome Reaction. *Biol Reprod* 41: 790–797.

Zorn, B., J. Auger, V. Velikonja, M. Kolbezen, and H. Meden-Vrtovec. 2008. Psychological Factors in Male Partners of Infertile Couples: Relationship with Semen Quality and Early Miscarriage. *Int J Androl* 31: 557–564.

Zukerman, Z., L. J. Rodriguez-Rigau, K. D. Smith, and E. Steinberger. 1977. Frequency Distribution of Sperm Counts in Fertile and Infertile Males. *Fertil Steril* 28: 1310–1313.

INDEX